Advanced Telecommunications Engineering

Advanced Telecommunications Engineering

Edited by **Bernhard Ekman**

C WILLFORD PRESS

New York

Published by Willford Press,
118-35 Queens Blvd., Suite 400,
Forest Hills, NY 11375, USA
www.willfordpress.com

Advanced Telecommunications Engineering
Edited by Bernhard Ekman

International Standard Book Number: 978-1-68285-159-3 (Hardback)

Printed in the United States of America.

Contents

Preface

This book has been an outcome of determined endeavour from a group of educationists in the field. The primary objective was to involve a broad spectrum of professionals from diverse cultural background involved in the field for developing new researches. The book not only targets students but also scholars pursuing higher research for further enhancement of the theoretical and practical applications of the subject.

Telecommunications engineering is a multidisciplinary field of study which incorporates concepts from electrical, electronic, computer and structural engineering. It focuses on the designing and maintenance of telecommunication equipment and systems. This book is a compilation of chapters that provide an insight into the latest advancements in the field of telecommunications engineering. Mobile cellular networks, routing protocols and techniques, modelling and simulation of wireless and mobile networks, etc. are some of the important topics discussed herein. Students and researchers will find this book very useful in understanding the significant concepts, technological advancements and emerging trends in telecommunications engineering.

It was an honour to edit such a profound book and also a challenging task to compile and examine all the relevant data for accuracy and originality. I wish to acknowledge the efforts of the contributors for submitting such brilliant and diverse chapters in the field and for endlessly working for the completion of the book. Last, but not the least; I thank my family for being a constant source of support in all my research endeavours.

Editor

REQUIREMENTS OF VERTICAL HANDOFF MECHANISM IN 4G WIRELESS NETWORKS

Mandeep Kaur Gondara[1] and Dr. Sanjay Kadam[2]

[1] Ph. D Student, Computer Science Department, University of Pune, Pune
u08401@cs.unipune.ac.in

[2] Research Guide, Computer Science Department, University of Pune, Pune
sskadam@cdac.in

ABSTRACT

The importance of wireless communication is increasing day by day throughout the world due to cellular and broadband technologies. Everyone around the world would like to be connected seamlessly anytime anywhere through the best network. The 4G wireless system must have the capability to provide high data transfer rates, quality of services and seamless mobility. In 4G, there are a large variety of heterogeneous networks. The users for variety of applications would like to utilize heterogeneous networks on the basis of their preferences such as real time, high availability and high bandwidth. When connections have to switch between heterogeneous networks for performance and high availability reasons, seamless vertical handoff is necessary. The requirements like capability of the network, handoff latency, network cost, network conditions, power consumption and user's preferences must be taken into consideration during vertical handoff. In this paper, we have extracted the requirements of a vertical handoff from the literature surveyed. The evaluation of the existing work is also being done on the basis of required parameters for vertical handoff. A sophisticated, adaptive and intelligent approach is required to implement the vertical handoff mechanism in 4G wireless networks to produce an effective service for the user by considering dynamic and non dynamic parameters.

KEYWORDS

4G wireless networks, VHO, Requirements, RSS, Parameters, Performance

1. INTRODUCTION

Mobility is the most important feature of today's wireless networking system. Mobility can be attained by handoff mechanisms in wireless networks. Handoff is the process of changing the channel (frequency, time slot, spreading code, or combination of them) associated with the current connection while a call is in progress [1].

1.1 Types of Handoffs in 4G Networks

In 4G networks, the handoffs are classified into two main streams

1.1.1 Horizontal Handoff

Handoff between two base stations (BSs) of the same system is called Horizontal handoff. Horizontal handoff involves a terminal device to change cells within the same type of network (e.g., within a CDMA network) to maintain service continuity [2]. It can be further classified into Link-layer handoff and Intra-system handoff. Horizontal handoff between two BS, under same foreign agent (FA) is known as Link-layer handoff. In Intra-system handoff, the horizontal handoff occurs between two BSs that belong to two different FAs and both FAs belongs to the same system and hence to same gateway foreign agent (GFA).

1.1.2   Vertical Handoff (VHO)

Vertical handoff refers to a network node changing the type of connectivity it uses to access a supporting infrastructure, usually to support node mobility. For example, a suitably equipped laptop might be able to use both a high speed wireless LAN and a cellular technology for Internet access. Wireless LAN connections generally provide higher speeds, while cellular technologies generally provide more ubiquitous coverage. Thus the laptop user might want to use a wireless LAN connection whenever one is available, and to 'fall over' to a cellular connection when the wireless LAN is unavailable. Vertical handovers refer to the automatic fallover from one technology to another in order to maintain communication [3].The vertical handoff mechanism allows a terminal device to change networks between different types of networks (e.g., between 3G and 4G networks) in a way that is completely transparent to end user applications[2].

The vertical handoff process involves three main phases [4], [5], namely system discovery, vertical handoff decision, and VHO execution. During the system discovery phase, the mobile terminal determines which networks can be used. These networks may also advertise the supported data rates and Quality of Service (QoS) parameters. In VHO decision phase, the mobile terminal determines whether the connections should continue using the current network or be switched to another network. The decision may depend on various parameters or metrics including the type of the application (e.g., conversational, streaming), minimum bandwidth and delay required by the application, access cost; transmit power, and the user's preferences. During the VHO execution phase, the connections in the mobile terminal are re-routed from the existing network to the new network in a seamless manner. This phase also includes the authentication, authorization, and transfer of a user's context information [6].

Handoff management aims at controlling the change of an access point (AP) in order to maintain the connection with the moving device during the active data transmission. The problem is exacerbated by the presence of APs adopting different technologies. Hence vertical handoffs, that is, handoff procedures between APs of heterogeneous technology, should be taken into account [7].

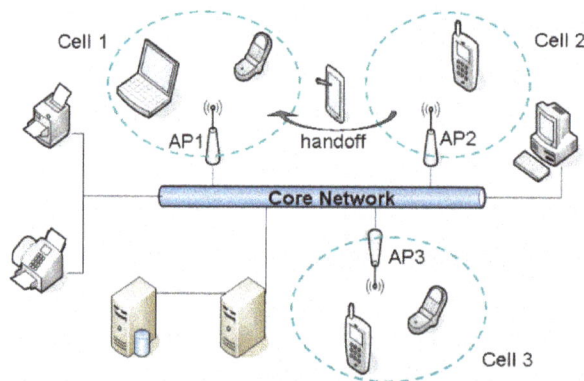

Figure 1: Vertical Handoff in heterogeneous networks

In this paper, we propose requirements for vertical handoff decision model for heterogeneous 4G networks on the basis of literature surveyed. Section 4 represents dynamic and non dynamic parameters for VHO (vertical handover) mechanism in 4G networks. In section 5, the evaluation of existing work is being done on the basis of parameters discussed in section4.

2. STATE OF THE ART

From the literature surveyed, different authors use different terms such as models, techniques and approaches to refer mechanisms. In order to compile a VHO mechanism for 4G wireless networks, it is essential to study existing VHO mechanisms. The study of existing mechanisms will assist in the identification of requirements for VHO mechanism. As of now, a few approaches for VHO have been found in the literature. One kind of approach is based on "Received Signal Strength (RSS)" that may be combined with other parameters such as network load and network cost. The another kind of approaches are using artificial intelligence techniques, combining several parameters such as network conditions and Mobile Terminal's (MT) mobility in the handoff decision[9]. Some are policy based approaches, combining several metrics such as access cost, power consumption, and bandwidth, velocity of a host, quality of service in VHO mechanism.

. Wang et al. [8] have introduced the policy enabled handoff (PEHHWN). In their work, they describe a policy-enabled handoff system that allows users to express policies and to find out the best network on the basis of dynamic and static parameters such as network cost, performance and power consumption. However, the cost function presented in that paper is very preliminary and cannot handle sophisticated configurations. Another policy based work is proposed in [11], where the Automatic Handover Manager (AHM) provides a solution for determining the best network interfaces for the services (TAHDM). The decision is made by using the context information from the mobile node, networks and the user as well as the RSS. AHM is based on the autonomic computing concept. It provides a good policy for the vertical handover using the context information without user's interventions. AHM has four major functions such as monitoring, analyzing, planning and executing. This paper describes how to compose the context evaluation function and formulate a policy. In future work, more concrete context information and improvements can be made in AHM by optimizing the context evaluation function. In [16], the objective of research work is to determine the conditions under which vertical handoff should be performed for heterogeneous wireless networks (VHDAHWN). This work incorporated the connection duration and signalling load incurred on the network for VHO decision. The work is based on the Markov decision process (MDP) formulation to maximize the expected total reward of a connection. Numerical results show that their proposed MDP algorithm gives a higher expected total reward and lower expected number of vertical handoffs than SAW (Simple Additive Weighting) and GRA (Grey Relational Analysis), and two heuristic policies under a wide range of conditions. Their proposed model is adaptive and applicable to a wide range of conditions.

A. Dvir et al. [13] proposed an efficient decision handoff mechanism for heterogeneous network (EDHMHN). A decision function in which the system considers all the available network and user parameters (e.g. host velocity, battery status, Wi-Fi AP's current load, and WiMAX BS's Qos guaranties, and performs technology selection such that an overall system performance metric is optimized(i.e., throughput and capacity limitation). They have defined a new system-wise entity that is activated when a user is in an area with overlapping access technologies and needs to decide the best technology to be used, where the entity performs technology selection in order to optimize the overall system performance metric in terms of throughput and capacity limitation. Their simulation results validate the efficiency of their method and show that it is also applicable to other combinations of access technologies.

Mrs. Chandralekha et al. [12] proposed a theory for selection of the best available wireless network during handoffs based on a set of predefined user preferences on a mobile device (UARTVHO). A neural network model has been used to process multi-criteria VHO decision metrics. The features used from generated data have been carefully selected and used as inputs for the neural network in order to have high performance rate. A modified type of competitive learning called "Adaptive Resonance Theory (ART)" has been designed to overcome the problem of learning stability. The proposed method is capable of selecting the best available wireless network with a reasonable performance rate. The overall approach is based on artificial intelligence, combining some other metrics for decision model of VHO.

Goyal et al. [9] proposed a dynamic decision model for VHO across heterogeneous wireless networks (ADDMVHO). This model makes the right VHO decisions by determining the "best" network at "best" time among available networks based on dynamic factors such as RSS and velocity of mobile station as well as static factors. A handoff Management Center (HMC) monitors the various inputs collected from the network interfaces and their base stations (BS) analyze this information and make handoff decisions. The dynamic algorithm has different phases. The Priority Phase is used to remove all the unwanted and ineligible networks from the prospective candidate networks. The Normal Phase is used to accommodate user-specific preferences regarding the usage of network interfaces. Finally, the Decision Phase is used to select the "Best" network and executing the handoff to the selected network. In [10], the proposed research provides optimized performance in heterogeneous wireless networks during VHO decision (VHDAPOP). A VHO decision algorithm is being developed that enables a wireless access network to balance the overall load among all attachment points (e.g., Base Stations (BSs) and Access Points (APs) and also to maximize the collective battery lifetime of mobile nodes (MNs). In addition, when ad hoc mode is applied to 3/4G wireless data networks, a route selection algorithm has been devised to forward data packets to the most appropriate attachment point for maximizing the collective battery lifetime and to maintain load balancing.

3. SOURCES OF SURVEY

The requirements of a VHO mechanism for wireless networks extracted from the literature surveyed. In the research, we used documentary sources, which involve existing textual documents available in electronic and printed media. The data sources used in this research include academic journals, applicable books and the internet.

In this survey, textual analysis is used as means of data collection. Textual analysis involves both content analysis and textual interpretations. Based on the research departure points, namely the vertical handoff mechanism and parameters of handoffs in wireless networks, the contents of the referenced publications were analyzed to find their applicability to the study. The requirements were extracted from existing handoff decision models and mechanisms and other surveyed literature on handoff aspects in the wireless network environment.Different authors have indicated different aspects that should be considered while designing a handoff mechanism for the wireless networks. These aspects include many parameters as discussed in section 4 for seamless and secure handoffs in wireless environment.

4. REQUIREMENT FOR HANDOFF MECHANISM

4.1 Bandwidth

Bandwidth is a measure of the width of a range of frequencies. It is the difference between the upper and lower frequencies in a contiguous set of frequencies. In order to provide seamless handoff for Quality of service (Qos) in wireless environment, there is a need to manage bandwidth requirement of mobile node during movement. Bandwidth is generally known as the link capacity in a network. Higher offered bandwidth ensures lower call dropping and call blocking probabilities; hence higher throughput [9].Bandwidth handling should be an integral part of any of the handoff technique.

4.2 Handoff Latency

Handover of calls between two BS is encountered frequently and the delay can occur during the process of handoffs. This delay is known as handoff latency. A good handoff decision model should consider Handoff latency factor and the handoff latency should be minimized. Many proposed handoff decision models have tried to minimize the handoff latency by incorporating this factor in their handoff decision models. Handoff Latencies affect the service quality of many applications of mobile users. It is essential to consider handoff latency while designing any handoff technique.

4.3 Power Consumption

In 4G networks, we need to find ways to improve energy efficiency. Power is not only consumed by user terminal but also attributed to base station equipments. Power is also consumed during mobile switching or handoffs. During handoff, frequent interface activation can cause considerable battery drainage. The issue of power saving also arises in network discovery because unnecessary interface activation can increase power consumption. It is also important to incorporate power consumption factor during handoff decision.

4.4 Network Cost

A multi criteria algorithm for handoff should also consider the network cost factor. The cost is to be minimized during VHO in wireless networks. The new call arrival rates and handoff call arrival rates can be analyzed using cost function. Next Generation heterogeneous networks can combine their respective advantages on coverage and data rates, offering a high Quality of Service (QoS) to mobile users. In such environment, multi-interface terminals should seamlessly switch from one network to another in order to obtain improved performance or at least to maintain a continuous wireless connection. Therefore, network selection cost is important in handoff decisions.

4.5 User Preferences

When handover happens, the users have more options for heterogeneous networks according to their preferences and network performance parameters. The user preferences could be preferred networks, user application requirements (real time, non-real time), service types (Voice, data, video), Quality of service (It is a set of technologies for managing network traffic in a cost effective manner to enhance user experiences for wireless environments) etc. User Preferences can also be considered for VHO in 4G wireless networks.

4.6 Network Throughput

Network throughput refers to the average data rate of successful data or message delivery over a specific communications link. Network throughput is measured in bits per second (bps). Maximum network throughput equals the TCP window size divided by the round-trip time of communications data packets. As network throughput is considered in dynamic metrics for making decision of VHO, it is one the important requirement to be considered for the VHO.

4.7 Network Load Balancing

Network load is to be considered during effective handoff. It is important to balance the network load to avoid deterioration in quality of services. Variations in the traffic loads among cells will reduce the traffic-carrying capacity. To provide a high quality communication service for mobile subscribers and to enhance a high traffic-carrying capacity when there are variations in traffic, network load must be paid attention.

4.8 Network Security

With the increasing demand of wireless networks, seamless and secure handoff has become an important factor in wireless networks. The network security consists of the provisions and policies adopted by the network to prevent and monitor unauthorized access, misuse, modification, and network-accessible resources. In a wireless environment, data is broadcast through the air and people do not have physical controls over the boundaries of transmissions. The security features provided in some wireless products may be weaker; to attain the highest levels of integrity, authentication, and confidentiality, network security features should be embedded in the handoff policies.

4.9 Received Signal strength (RSS)

The performance of a wireless network connection depends in part on signal strength. Between a mobile node (MN) and access point (AP), the wireless signal strength in each direction determines the total amount of network bandwidth available along that connection. RSS depicts the power present in a received signal. A signal must be strong enough between base station and mobile unit to maintain signal quality at receiver. The RSS should not be below a certain threshold in a network during handoff. VHO includes three sequential steps as discussed earlier in this paper, namely handoff initiation, handoff decision and handoff execution. Handoff initiation is concerned with measurement of RSS [14].

4.10   Velocity

Velocity of the host should also be considered during handoff decision. Because of the overlaid architecture of heterogeneous networks, handing off to an embedded network, having small cell area, when travelling at high speeds is discouraged since a handoff back to the original network would occur very shortly afterwards [9].

However, we have stated the important parameters/metrics as requirements but other parameters such as network conditions, network capability and bit error rate can also be considered during vertical handoff. The dynamic requirements include RSS, velocity, throughput, user preferences as parameters and non-dynamic requirements include network cost, power consumption, network security and bandwidth as parameters. A good handoff mechanism decision model should have both dynamic and non-dynamic metrics. However, it is important to consider maximum number of static and dynamic requirements during VHO but it is difficult to include all the metrics in a single decision model due to complexity of algorithms and conflicting issues of multiple metrics.

5. EVALUATION OF THE EXISTING WORK

In this section, existing handoff mechanisms are evaluated against the requirements of a VHO in a wireless environment. The evaluation is intended to establish the gap between the existing handoff mechanisms and the handoff requirements for the wireless environment. In table 1 we are listing the handoff models for reference purpose in table 2. In table2, the evaluation of existing work is done against the requirements for VHO discussed in section4.

Abbreviation	Ref#	Handoff Mechanisms
UARTVHO	12	Use of Adaptive Resonance Theory for Vertical Handoff Decision in heterogeneous Wireless Environment
VHDAHWN	16	A Vertical Handoff Decision Algorithm for Heterogeneous Wireless Networks
TAHDM	11	Towards Autonomic Handover Decision Management in 4G Networks
EDHMHN	13	Efficient decision handoff mechanism for heterogeneous network
PEHHWN	8	Policy-Enabled Handoffs across Heterogeneous Wireless Networks
VHDAPOP	10	Vertical Handoff Decision Algorithms for Providing Optimized Performance in heterogeneous Wireless Networks
ADDMVHO	9	A Dynamic Decision Model for Vertical Handoffs across Heterogeneous Wireless Networks

Table 1: Abbreviations for the existing handoff mechanisms

Handoff Mechanism	Requirement Number									
	1	2	3	4	5	6	7	8	9	10
UARTVHO	X	X		X	X	X	X	X		
VHDAHWN	X			X	X	X	X		X	
TAHDM	X		X	X	X	X			X	X
EDHMHN	X		X		X	X	X		X	X
PEHHWN	X	X	X	X			X			
VHDAPOP	X		X	X	X		X			
ADDMVHO	X		X	X					X	X

Table 2: Requirements satisfied by the existing handoff mechanism

Using the requirement parameters as stated in section 4, it has been found that there is no vertical handoff mechanism that satisfies all the requirements, although all VHO models/mechanisms satisfy at least five requirements.

- For requirement 1, vertical handoff mechanisms discussed above indicate the importance of bandwidth in wireless networks. However, Vertical handoff models such as UARTVHO, VHDAHWN, EDHMHN, PEHHWN, VHDAPOP, and ADDMVHO have incorporated this requirement in their model but still a mechanism is required for controlling the variations in bandwidth while a MN is switching from high to low or low to high bandwidth network.

- For Requirement 2, the delay during the handover process is to be minimized. Handoff mechanisms such as UARTVHO and VHDAHWN have tried to minimize the handoff latency by incorporating this factor in their handoff decision models.

- For requirement 3, there is a need to find ways to improve energy efficiency. The handoff models that have considered this requirement are TAHDM, EDHMHN, VHDAPOP, PEHHWN and ADDMVHO.

- For requirement 4, the network selection cost is to be incorporated in the decision model of a VHO. These includes UARTVHO, VHDAHWN, TAHDM, VHDAPOP, PEHHWN and ADDMVHO mechanisms which are using cost functions to analyze the network cost during switching of networks by a mobile node.

- For requirement 5, the user preferences must be considered in terms of preferred network, service types and requirements of applications. The frameworks that consider user preferences during handoff decision are UARTVHO, VHDAHWN, TAHDM, VHDAPOP and PEHHWN.

- Requirement 6 states that the VHO mechanism should consider average data rate of successful data over a communication link. Handoff mechanisms such as UARTVHO, VHDAHWN, TAHDM and EDHMHN are able to satisfy this requirement.

- For requirements 7, it is important to balance the network load for traffic carrying capacity, quality of services and for providing high quality communication. Many of the proposed handoff mechanisms satisfy this requirement such as VHDAHWN, VHDAPOP, UARTVHO, EDHMHN, and PEHHWN.

- For requirement 8, secure handoff has become an important factor in wireless networks. Network security feature must be merged along with other parameters in the decision model of VHO. As in table 2, only one work "UARTVHO" considers this parameter. There is a need to integrate network security during handoff.

- Requirement 9 states that the signal strength plays crucial role in performance of the wireless network by depicting the power present in a signal. The state of the art include VHDAHWN,TAHDM, EDHMHN and ADDMVHO work where handoff decision model is based on the RSS.

- For requirement 10, velocity of the host must be paid attention during handoff decision. The handoff mechanisms such as EDHMHN, PEHHWN and ADDMVHO consider this requirement.

However, the evaluation of the existing work is being done on the basis of requirements extracted from the literature surveyed but the evaluation is not being done on the basis of performance of the algorithms. There are basically three mechanisms used to evaluate the performance of the handoff mechanisms, analytical, simulation and emulation approaches as presented in the literature. The analytical approach works well under certain specified constraints. The simulation approach is used commonly to verify the experimental results on the simulator. The emulation approach uses a software simulator consisting of handoff algorithms to process measured variables (for instance, received signal strength and bit error rate) [17]. Before deciding the parameters for decision mechanism of VHO, the performance based evaluation can be done for better results.

The decision mechanism of VHO can become more fruitful, if the number of parameters is more during decision making as stated in section 4. The success of vertical handoff mechanism depends upon the decision model based on requirements/metrics. An efficient VHO decision mechanism can not only enhance the system capacity but also improve the quality of services for a user. The existing works can be extended or new works can be developed to incorporate more parameters in VHO mechanism.

6. CONCLUSION

The vertical handoff will remain an essential component for 4G wireless networks due to switching of mobile users amongst heterogeneous networks. In this paper, we have described a few works in vertical handoff mechanisms and exposed a summary of decision algorithms for vertical handoff in literature. The 4G wireless networks create new handoff challenges due to multiple requirements for vertical handoff. In this paper, the requirements of a vertical handoff for 4G wireless network were proposed. The requirements include high bandwidth, low handoff latency, lower power consumption, minimum network cost, balanced network load, network security, user preferences, throughput and RSS of a switching network. Establishing the requirements of a vertical handoff mechanism for 4G wireless networks is a critical milestone in the development of vertical handoff mechanism for 4G. In this paper, the evaluation of existing vertical handoff mechanisms is also done against the requirements stated in the paper. The evaluation indicate the need to have a VHO mechanism for 4G wireless networks that has the ability to satisfy maximum number of requirements. However, it is difficult to consider all the parameters during designing the decision model for VHO but if we consider more parameters, the outcome of the decision mechanism would definitely improve.

7. REFERENCES

[1]Qing-An Zeng & Dharma P. Agrawal, (2002) "Handbook of Wireless Networks and Mobile Computing", John Wiley & Sons Publishers.

[2]James Won-ki Hong & Alberto Leon-Garcia, (2005) "Requirements for the Operations and Management of 4G networks", In Proc. of 19th International Conference on Performance Challenges for Efficient Next Generation Networks, pp 981-990.

[3]http://www.tutorvista.com/-United States

[4] J. McNair & F. Zhu, June (2004) "Vertical Handoffs in Fourth-generation Multi-network Environments", IEEE Wireless Communications, Vol. 11, No. 3, pp 8–15.

[5] W. Chen & Y. Shu, March (2005) "Active Application Oriented Vertical Handoff in Next Generation Wireless Networks", In Proc. of IEEE WCNC'05, New Orleans, LA.

[6]Enrique Stevens-Navarro, Vincent W.S. Wong & Yuxia Lin,March (2007) "A Vertical Handoff Decision Algorithm for Heterogeneous Wireless Networks", In Proc. of IEEE Wireless Communications and Networking Conference (WCNC'07), Hong Kong, China.

[7] http://www.mobilab.unina.it/locationing.htm

[8]H.J. Wang, R. H. Katz & J. Giese,(1999) "Policy-Enabled Handoffs across Heterogeneous Wireless Networks", In proc. of ACM WMCSA.

[9]Pramod Goyal & S. K. Saxena,(2008)"A Dynamic Decision Model for Vertical Handoffs across Heterogeneous Wireless Networks",677 ? 2008 WASET.ORG, World Academy of Science, Engineering and Technology,Issue 41,pp 676-682.

[10]SuKyoung Lee , Kotikalapudi Sriram, Kyungsoo Kim, Yoon Hyuk Kim & Nada Golmie,January (2009)"Vertical Handoff Decision Algorithms for Providing Optimized Performance in Heterogeneous Wireless Networks",IEEE transactions on Vehicular Technology.

[11]Joon-Myung Kang, Hong-Taek Ju2 & James Won-Ki Hong,(2006) "Towards Autonomic Handover Decision Management in 4G Networks", In Proceedings of MMNS'2006,pp145-157.

[12]Mrs. Chandralekha & Dr. Prafulla Kumar Behera,November (2009) "Use of Adaptive Resonance Theory for Vertical Handoff Decision in Heterogeneous Wireless Environment", International Journal of Recent Trends in Engineering, Vol. 2, No. 3.

[13]A. Dvir, R. Giladi, I. Kitroser & M. Segal,February (2010) "Efficient decision handoff mechanism for heterogeneous network", International journal of Wireless and Mobile networks,Vol. 2, No. 1.

[14]K.Ayyappan and P.Dananjayan,(2008)"RSS Measurement for Vertical Handoff in Hetrogeneous Network", International journal of Theoretical and Applied Information Technology,Vol.4,No.10.

[15]KunHo Hong,SuKyoung Lee, LaeYoung Kim & PyungJung Song,(2009) "Cost-Based Vertical Handover Decision Algorithm for WWAN/WLAN Integrated Networks",EURASIP Journal on Wireless Communications and Networking Volume 2009 , Article ID 372185, 11 pages doi:10.1155/2009/372185.

[16]E.Stevens-Navarro,Vincent W.S.Wong & Yuxia Lin,(2007) "A Vertical Handoff Decision Algorithm for Heterogeneous Wireless Networks",In Proc. of Wireless Communications and Networking Conference, IEEE ; doi:10.1109/WCNC. 2007.590

[17] A.J.Onumanyi & E.N.Onwuka,(2011) " Techniques for vertical handoff decision across wireless heterogeneous networks: A survey", in Academic Journal of Scientific Research and Essays,Vol. 6(4),pp.683-687.

2

Impact of the Optimum Routing and Least Overhead Routing Approaches on Minimum Hop Routes and Connected Dominating Sets in Mobile Ad hoc Networks

Natarajan Meghanathan

Jackson State University, 1400 Lynch St, Jackson, MS, USA
natarajan.meghanathan@jsums.edu

ABSTRACT

Communication protocols for mobile ad hoc networks (MANETs) follow either an Optimum Routing Approach (ORA) or the Least Overhead Routing Approach (LORA): With ORA, protocols tend to determine and use the optimal communication structure at every time instant; whereas with LORA, a protocol tends to use a chosen communication structure as long as it exists. In this paper, we study the impact of the ORA and LORA strategies on minimum hop routes and minimum connected dominating sets (MCDS) in MANETs. Our primary hypothesis is that the LORA strategy could yield routes with a larger time-averaged hop count and MCDS node size when compared to the minimum hop count of routes and the node size of the MCDS determined using the ORA strategy. Our secondary hypothesis is that the impact of ORA vs. LORA also depends on how long the communication structure is being used. Our hypotheses are evaluated using extensive simulations under diverse conditions of network density, node mobility and mobility models such as the Random Waypoint model, City Section model and the Manhattan model. In the case of minimum hop routes, which exist for relatively a much longer time compared to the MCDS, the hop count of routes maintained according to LORA, even though not dramatically high, is appreciably larger (6-12%) than those maintained according to ORA; on the other hand, the number of nodes constituting a MCDS maintained according to LORA is only at most 6% larger than the node size of a MCDS maintained under the ORA strategy.

KEYWORDS

Minimum hop routes, Minimum connected dominating sets, Optimum routing approach, Least overhead routing approach, Mobile ad hoc networks, Simulations

1. INTRODUCTION

A mobile ad hoc network (MANET) is a dynamic distributed system of wireless nodes that move independently of each other. Routes in MANETs are often multi-hop in nature due to the limited transmission range of the battery-operated wireless nodes. MANET routing protocols are of two types [1][2]: proactive and reactive. Proactive routing protocols determine routes between every pair of nodes in the network, irrespective of their requirement. Reactive or on-demand routing protocols determine routes between any pair of nodes only if data needs to be transferred between the two nodes and no route is known between the two nodes. Proactive routing protocols always tend to maintain optimum routes between every source-destination (*s-d*) pair and this strategy is called the Optimum Routing Approach (ORA) [1][3]. In this pursuit, each node periodically exchanges its routing table and link state information with other nodes in the network, thus generating a significantly larger control overhead. On the other hand, reactive routing protocols use a Least Overhead Routing Approach (LORA) [1][3] wherein an *s-d* route is discovered through a global broadcast flooding-based route discovery process and the discovered route is used as long as it exists. With node mobility, an *s-d* route determined to be optimal at a particular time instant need not remain optimal in the subsequent time instants, even though the route may continue to exist. Thus, with LORA, it is possible that the routing protocols continue to send data packets through sub-optimal routes. On the other hand, with ORA, even though, we could send data packets at the best possible route at any time instant, the cost of periodically discovering such a route may be significantly high. In dynamically changing network topologies, reactive on-demand routing protocols have been preferred over proactive protocols with respect to the routing control overhead incurred [4][5].

From another perspective, among the routing algorithms and protocols proposed for MANETs, routing based on a connected dominating set (CDS) has been recognized as a suitable approach in adapting quickly to the unpredictable fast-changing topology and dynamic nature of a MANET [6]. It is considered adaptable because as long as topological changes do not affect the structure of the CDS, there is no need to reconfigure the CDS since the routing paths based on the CDS would still be valid. A MANET is often represented as a unit disk graph [7] built of vertices and edges, where vertices signify nodes and edges signify bi-directional links that exist between any two nodes if they are within each other's transmission range. In a given graph representing a MANET, a CDS is a dominating set within the graph whose induced sub graph is connected. A dominating set of a graph is a vertex subset, such that every vertex is either in the subset or adjacent to a vertex in the subset [8]. Routing based on a CDS within a MANET means that routing control messages will be exchanged only amongst the CDS nodes and not broadcast by all the nodes in the network; this will reduce the number of unnecessary transmissions in routing [9].

There are multiple ways to form a CDS within a given MANET, and the algorithm used for CDS formation will affect the performance and lifetime of the CDS and the performance of the MANET as a whole. A popular approach in CDS formation is attempting to form the smallest possible CDS within a MANET, referred to as a minimum connected dominating set (MCDS). Reducing the size of the CDS will mean reducing the number of unnecessary transmissions. Unfortunately, the problem of determining a MCDS in an undirected graph like that of the unit disk graph is NP-complete [9][12]. Efficient heuristics [10][11][12] have been proposed to approximate the MCDS in wireless ad hoc networks. A common thread among these heuristics is to give the preference of CDS inclusion to nodes that have high neighborhood density. The MaxD-CDS heuristic [9] that we study in this paper is one such heuristic.

The objective of this paper is to study the impact of adopting the ORA and LORA strategies on minimum hop routes and the node size of the MCDS in MANETs. Minimum hop routing is a very widely adopted route selection principle of MANET routing protocols, belonging to both proactive and reactive categories. Likewise, the primary objective of a majority of the MCDS-based heuristics is to minimize the number of nodes constituting the CDS. As ORA determines the best optimal route at any time instant, our primary hypothesis is that the hop count of minimum hop routes and the node size of MCDS discovered under the LORA strategy would be greater than those discovered under the ORA strategy. Our secondary hypothesis is that the impact of ORA vs. LORA also depends on how long the communication structure is being used. We determine the percentage difference in the hop count of minimum hop s-d paths and the node size of the MCDS determined under the two strategies. We conduct extensive simulations under three different network densities and three different mobility models with three different levels of node mobility. The three mobility models [13] used are the Random Waypoint model, City Section model and Manhattan model. Even though performance comparison studies of individual proactive vs. reactive routing protocols as well as the different CDS algorithms are available in the literature, an extensive simulation based analysis on the impact of the ORA and LORA strategies on the minimum hop count of routes and the node size of the MCDS algorithms has not been conducted in the literature and therein lies our contribution through this paper.

The rest of the paper is organized as follows: Section 2 discusses the algorithms employed for determining minimum hop routes under the ORA and LORA strategies and also illustrates an example highlighting the difference between the two strategies and their impact on the hop count of s-d paths. Section 3 discusses the algorithms employed for determining MCDS under the ORA and LORA strategies and also illustrates an example highlighting the difference between the two strategies and their impact on the node size of the MCDS. Section 4 reviews the three different mobility models used in the simulations. Section 5 describes the simulation environment and presents the simulation results for hop count per s-d path, node size per MCDS, path lifetime and network connectivity. Section 6 concludes the paper and lists future work. Throughout the paper, the terms 'node' and 'vertex', 'edge' and 'link', 'path' and 'route' are used interchangeably. They mean the same.

2. DETERMINATION OF MINIMUM HOP ROUTES UNDER THE ORA AND LORA STRATEGIES

We use the notion of a mobile graph [14] defined as the sequence $G_M = G_1G_2 ... G_T$ of static graphs that represent the network topology changes over the time scale T, representing the simulation time. We sample the network topology periodically, for every 0.25 seconds, which in reality could be the instants of data packet origination at the source. Each of the static graphs is a unit disk graph [7] of nodes and edges, wherein there exists an edge if and only if the Euclidean distance between the two constituent end nodes of the edge is within the transmission range of the nodes. We assume every node operates at a fixed transmission range, R.

For the ORA strategy, we determine the sequence of minimum hop s-d paths between a source node s and a destination node d by running the Breadth First Search (BFS) algorithm [15], starting from the source node s, on each of the static graphs of the mobile graph generated over the entire time period of the simulation. In the case of LORA, if we do not know a path from source s to destination d in static graph G_i, we run BFS (pseudo code in Figure 1), starting from node s, on G_i and determine the minimum hop path P_{s-d} from s to d. For subsequent static graphs G_{i+1}, G_{i+2}, ..., we simply test the presence of path P_{s-d}. We validate the existence of a path P_{s-d} in static graph G_j by testing the existence of every constituent edge of P_{s-d} in G_j. If every constituent edge of P_{s-d} exists in G_j, then the path P_{s-d} exists in G_j. Otherwise, we run BFS on G_j, starting from the source node s, and determine a new s-d path P_{s-d}. This procedure is repeated until the end of the simulation time. The pseudo code of our algorithms to determine the minimum hop paths under the ORA and LORA strategies is given in Figures 2 and 3 respectively.

Input: Static Graph $G = (V, E)$, source node s, destination node d
Auxiliary Variables/Initialization: *Nodes-Explored* = Φ, *FIFO-Queue* = Φ
$\qquad\qquad\qquad\qquad\qquad \forall$ node $v \in V$, *Parent* (v) = NULL
Begin Algorithm *BFS* (G, s, d)
$\quad$ *Nodes-Explored = Nodes-Explored* U $\{s\}$
$\quad$ *FIFO-Queue = FIFO-Queue* U $\{s\}$
$\quad$ **while** (|*FIFO-Queue*| > 0) **do**
$\qquad$ node u = Dequeue(*FIFO-Queue*) // extract the first node
$\qquad$ **for** (every edge (u, v)) **do** // i.e. every neighbor v of node u
$\qquad\quad$ **if** ($v \notin$ *Nodes-Explored*) **then**
$\qquad\qquad$ *Nodes-Explored = Nodes-Explored* U $\{v\}$
$\qquad\qquad$ *FIFO-Queue = FIFO-Queue* U $\{v\}$
$\qquad\qquad$ *Parent* $(v) = u$
$\qquad\quad$ **end if**
$\qquad$ **end for**

```
end while
if ( | Nodes-Explored | = | V | ) then
    Path P_{d-s} = {d}
    temp-node = d
    while (Parent (temp-node) != NULL) do
        P_{d-s} = P_{d-s} U {Parent (temp-node)}
        temp-node = Parent (temp-node)
    end while
    Path P_{s-d} = reverse(P_{d-s})
    return P_{s-d}
end if
else
    return NULL // no s-d path
end if
End Algorithm BFS
```

Figure 1: Breadth First Search (BFS) Algorithm to Determine Minimum Hop s-d Path

```
Input: G_M = G_1 G_2 ... G_T, source s, destination d
Auxiliary Variable: i, Path P_{s-d}
Initialization: i=1; P_{s-d} = NULL
Begin ORA-MinHopPaths
    while (i ≤ T) do
        Path P_{s-d} = BFS(G_i, s, d)
        i = i + 1
    end while
End ORA-MinHopPaths
```

Figure 2: Pseudo Code to Find a Sequence of Minimum Hop s-d Paths under the ORA Strategy

```
Input: G_M = G_1 G_2 ... G_T, source s, destination d
Auxiliary Variables: i, j, Path P_{s-d}
Initialization: i=1; j=1; P_{s-d} = NULL
Begin LORA-MinHopPaths
    while (i ≤ T) do

        if (P_{s-d} != NULL) then
            for every edge (u, v) in P_{s-d} do
                if ( (u, v) does not exist in G_i) then
                    P_{s-d} = NULL
                end if
            end for
        end if
        if (P_{s-d} = NULL) then
            Path P_{s-d} = BFS(G_i, s, d)
        end if
        i = i + 1
    end while
End LORA-MinHopPaths
```

Figure 3: Pseudo Code to Find Sequence of Minimum Hop s-d Paths under the LORA Strategy

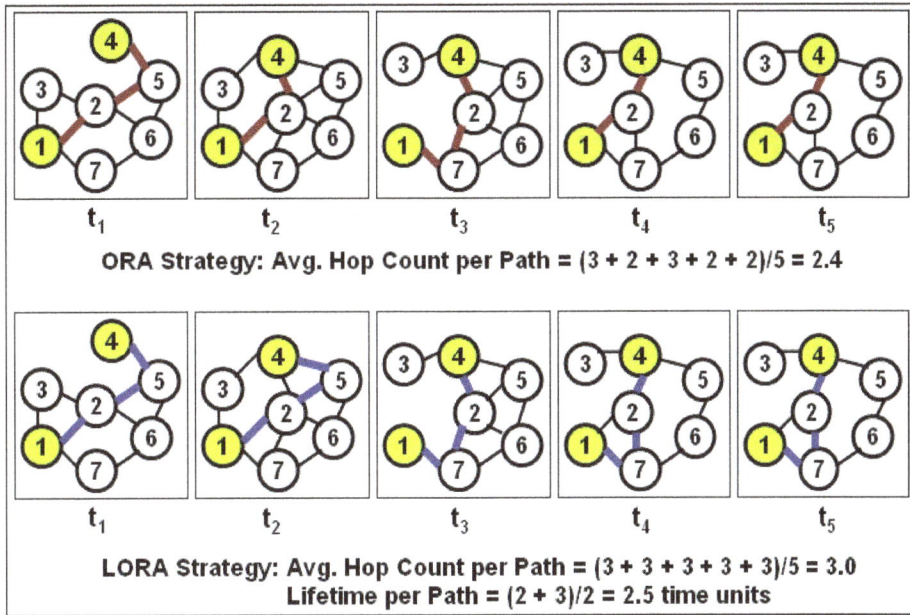

Figure 4: Example to Illustrate the ORA and LORA Strategies for Minimum Hop Routing

Figure 4 is an example to illustrate the difference between the ORA and LORA strategies with respect to minimum hop routing. We sample the network topology for five consecutive instants of time as shown. The source and destination node IDs are 1 and 4 respectively. We notice that under the LORA strategy, we could use path $\{1 - 2 - 4 - 5\}$ for time instants t_1 and t_2 and path $\{1 - 7 - 2 - 4\}$ for time instants t_3, t_4 and t_5 respectively. The paths $\{1 - 2 - 4 - 5\}$ and $\{1 - 7 - 2 - 4\}$ appear to be the best possible minimum hop paths at the time of discovery, i.e., at time instants t_1 and t_3 respectively. Nevertheless, after each of these paths is chosen at a particular time instant, we notice the emergence of relatively shorter paths (i.e., with a lower hop count) in the static graphs captured at subsequent time instants. But it is not possible to use these paths under the LORA strategy. With ORA, the strategy is to capture the minimum hop paths at every time instant.

3. DETERMINATION OF MINIMUM CONNECTED DOMINATING SETS UNDER THE ORA AND LORA STRATEGIES

The algorithm used to approximate a MCDS is referred to as the MaxD-CDS algorithm [9] as it prefers to include nodes that have a larger number of uncovered neighbors (density) to be part of the CDS. The MaxD-CDS algorithm uses the following principal data structures:

(i) *CDS-Node-List* – includes all nodes that are members of the CDS
(ii) *Covered-Nodes-List* – includes all nodes that are in the CDS-Node-List and all nodes that are adjacent to at least one member of the CDS-Node-List.

Before we run the MaxD-CDS algorithm, we make sure the underlying network graph is connected by running the Breadth First Search (BFS) algorithm [15]; because, if the underlying network graph is not connected, we would not be able to find a CDS that will cover all the nodes in the network. We run BFS, starting with an arbitrarily chosen node in the network graph. If we are able to visit all the vertices in the graph, then the corresponding network is said to be connected. If the graph is not connected, we simply collect a snapshot of the network topology at the next time instant and start with the BFS test. The pseudo code for the BFS algorithm is given in Figure 5.

Input: Graph $G = (V, E)$
Auxiliary Variables/Initialization: *Nodes-Explored* = Φ, *FIFO-Queue* = Φ
Begin Algorithm *BFS (G, s)*
   root-node = randomly chosen vertex in *V*
   Nodes-Explored = Nodes-Explored U {*root-node*}
   FIFO-Queue = FIFO-Queue U {*root-node*}
   while (|*FIFO-Queue*| > 0) **do**
     front-node *u* = Dequeue(*FIFO-Queue*) // extract the first node
     for (every edge (*u, v*)) **do** // i.e. every neighbor *v* of node *u*
       if (*v* ∉ *Nodes-Explored*) **then**
         Nodes-Explored = Nodes-Explored U {*v*}
         FIFO-Queue = FIFO-Queue U {*v*}
         Parent (*v*) = *u*
       end if
     end for
   end while

   if (| *Nodes-Explored* | = | *V* |) **then return** Connected Graph - true
   else return Connected Graph - false
   end if
End Algorithm BFS

Figure 5: Modified BFS Algorithm to Test for Graph Connectivity

The MaxD-CDS algorithm (pseudo code in Figure 6) outputs a *CDS-Node-List* based on a given input MANET graph. The first node to be included in the *CDS-Node-List* is the node with the maximum number of uncovered neighbors (any ties are broken arbitrarily). A CDS member is considered to be "covered", so a CDS member is additionally added to the *Covered-Nodes-List* as it is added to the *CDS-Node-List*. All nodes that are adjacent to a CDS member are also said to be covered, so the uncovered neighbors of a CDS member are also added to the *Covered-Nodes-List* as the member is added to the *CDS-Node-List*. To determine the next node to be added to the *CDS-Node-List*, we must select the node with the largest density amongst the nodes that meet the criteria for inclusion into the CDS. The criteria for CDS membership selection are the following: the node cannot already be a part of the CDS (*CDS-Node-List),* the node must be in the *Covered-Nodes-List*, and the node must have at least one uncovered neighbor (at least one neighbor that is not in the *Covered-Nodes-List*). Amongst the nodes that meet these criteria for CDS membership inclusion, we select the node with the largest density (i.e., the largest number of uncovered neighbors) to be the next member of the CDS. Ties are broken arbitrarily. This process is repeated until all nodes in the network are included in the *Covered-Nodes-List*. Once all nodes in the network are considered to be "covered", the CDS has been formed and the algorithm returns a list of the members included in the resultant MaxD-CDS (nodes in the *CDS-Node-List*).

Input: Graph $G = (V, E)$; *V* – vertex set, *E* – edge set
        Source vertex, *s* – vertex with the largest number of uncovered neighbors in *V*
Auxiliary Variables and Functions: *CDS-Node-List, Covered-Nodes-List, Neighbors*(*v*) for every *v* in *V*
Output: *CDS-Node-List*
Initialization: *Covered-Nodes-List* = {*s*}, *CDS-Node-List* = Φ
Begin Construction of *MaxD-CDS (G, s)*
   while (|*Covered-Nodes-List*| < |*V*|) **do**

Select a vertex $r \in$ *Covered-Nodes-List* and $r \notin$ *CDS-Node-List* such that r has the largest number of uncovered neighbors that are not in *Covered-Nodes-List*

    CDS-Node-List = *CDS-Node-List* U $\{r\}$

        for all $u \in$ *Neighbors*(r) and $u \notin$ *Covered-Nodes-List*
           Covered-Nodes-List = *Covered-Nodes-List* U $\{u\}$
        end for
      end while
    return *CDS-Node-List*
End Construction of *MaxD-CDS*

Figure 6: Pseudo Code for the Algorithm to Construct Maximum Density (MaxD)-based CDS

For the ORA strategy, we determine the sequence of MCDS by running the MaxD-CDS algorithm, starting from the source node s – the node with the largest number of neighbors, on each of the static graphs of the mobile graph generated over the entire time period of the simulation. In the case of LORA, if we do not know a MCDS in static graph G_i, we run the MaxD-CDS algorithm, starting from the source node s – the node with the largest number of neighbors in G_i and determine the MCDS. For subsequent static graphs G_{i+1}, G_{i+2}, ..., we simply test the presence of the MCDS. We validate a MCDS in a static graph G_j by first testing the connectivity among the nodes that constitute the MCDS and then testing whether each non-MCDS node in G_j is a neighbor of at least one node in the MCDS. If both these tests return true, then we consider the MCDS to exist in G_j. Otherwise, we run the MaxD-CDS algorithm on G_j, starting from a source node s – the node with the largest number of neighbors in G_j and determine a new MCDS. This procedure is repeated until the end of the simulation time. A pseudo code for the algorithm to validate a MCDS is given in Figure 7. The pseudo code of our algorithms to determine the MCDS under the ORA and LORA strategies is given in Figures 8 and 9 respectively.

Input: *CDS-Node-List* // Set of vertices part of the CDS
Auxiliary Variables and Functions:
    CDS-Edge-List – Set of edges, $\subseteq E$, between the vertices that are part of *CDS-Node-List*
    connectedCDS – Boolean variable that stores information whether *CDS-Node-List* and
                  CDS-Edge-List form a connected sub graph of G.
Output: *true* or *false*
      // *true*, if the nodes in *CDS-Node-List* form a connected sub graph of G and every vertex
           $v \notin$ *CDS-Node-List* is a neighbor of a vertex $u \in$ *CDS-Node-List*
      // *false*, if the nodes in *CDS-Node-List* do not form a connected sub graph of G and/or
        there exists at least one vertex $v \notin$ *CDS-Node-List* that has no neighbor in *CDS-Node-List*
Initialization: *CDS-Edge-List* = Φ
Begin CDS-Validation (*CDS-Node-List*, time instant t)
  for every pair of vertices $u, v \in$ *CDS-Node-List* **do**
    if there exists an edge $(u, v) \in E$ at time instant t **then**
      CDS-Edge-List = *CDS-Edge-List* U $\{(u, v)\}$
    end if
  end for
  connectedCDS = Breadth-First-Search(*CDS-Node-List*, *CDS-Edge-List*)
  if *connectedCDS* = true **then**

    for every vertex $v \notin$ *CDS-Node-List* **do**
      if there exists no edge $(u, v) \in E$ where $u \in$ *CDS-Node-List* at time instant t **then**

        return *false*
      end if
    end for

    return *true*

  end if

  return *false* // if *connectedCDS* = false

End CDS-Validation

Figure 7: Pseudo Code for the CDS Validation Algorithm

Input: $G_M = G_1G_2 \dots G_T$
Auxiliary Variable: i, $MCDS_i$
Initialization: $i=1$; $MCDS_i$ = NULL

Begin *ORA-MCDS*
  while $(i \leq T)$ **do**
    Choose the source node s – the node with the largest number of neighbors in G_i
    if *BFS* (G_i, s) returns **true**
      $MCDS_i$ = MaxD-CDS(G_i, s)
    end if
    $i = i + 1$
  end while
End *ORA-MCDS*

Figure 8: Pseudo Code to Determine a Sequence of MCDS under the ORA Strategy

---

Input: $G_M = G_1G_2 \dots G_T$, source s, destination d
Auxiliary Variables: i, j, $MCDS$
Initialization: $i=1$; $j=1$; $MCDS$ = NULL
Begin *LORA-MCDS*
  while $(i \leq T)$ **do**
    if $(MCDS$!= NULL$)$ **then**
      if (CDS-Validation (*CDS-Node-List* of *MCDS*, time instant i) returns **false**) **then**
        $MCDS$ = NULL
      end if
    end if
    if $(MCDS$ = NULL$)$ **then**
      Choose the source node s – the node with the largest number of neighbors in G_i
      if *BFS* (G_i, s) returns **true**
        $MCDS$ = MaxD-CDS(G_i, s)
      end if
    end if
    $i = i + 1$
  end while
End *LORA-MCDS*

Figure 9: Pseudo Code to Determine Minimum Hop *s-d* Paths under the LORA Strategy

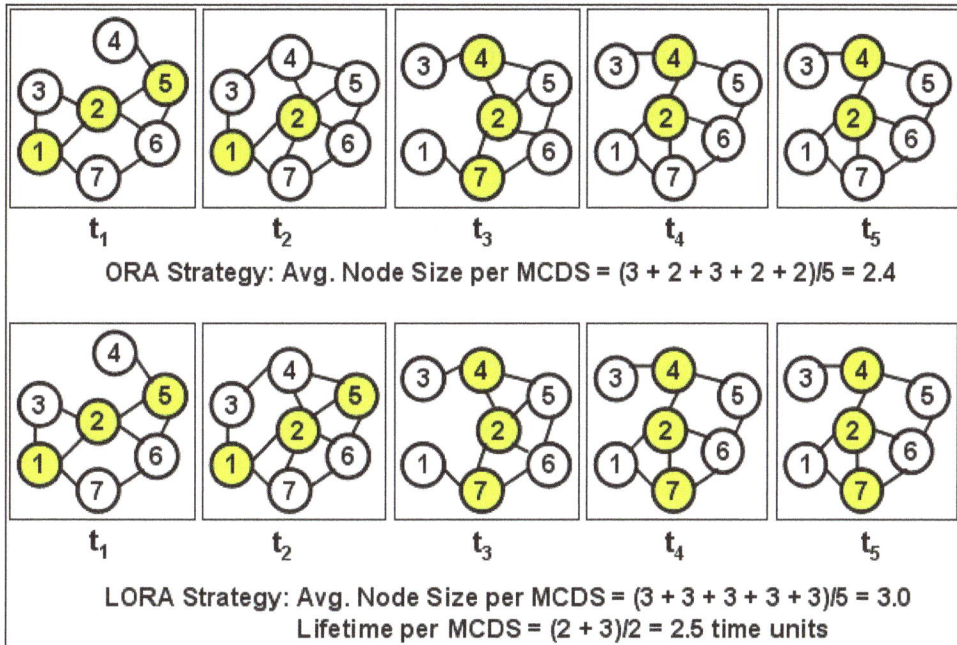

Figure 10: Example to Illustrate the ORA and LORA Strategies for Determining MCDS

Figure 10 is an example to illustrate the difference between the ORA and LORA strategies with respect to determining MCDS. We sample the network topology for five consecutive instants of time as shown. To determine a MCDS on a particular network topology, we start with the vertex (node ID 2 in all the cases) that has the largest number of neighbors. While deciding whether a covered node can be part of the *CDS-Node-List* of the MCDS, we include the covered node with the largest number of uncovered neighbors. Any tie in this case is broken in favor of the covered node that has the lowest ID. We notice that under the LORA strategy, we could use the MCDS comprising of nodes $\{2, 1, 5\}$ with edges $\{1 - 2, 2 - 5\}$ for time instants t_1 and t_2 and the MCDS comprising of nodes $\{2, 4, 7\}$ with edges $\{2 - 4, 2 - 7\}$ for time instants t_3, t_4 and t_5 respectively. The average MCDS node size under the LORA approach is 3.0 as there are three nodes in the MCDS used in each of the five time instants. On the other hand, under the ORA strategy, we determine MCDS comprising of nodes $\{2, 1, 5\}$, $\{2, 1\}$, $\{2, 4, 7\}$, $\{2, 4\}$, $\{2, 4\}$ at time instants t_1, t_2, t_3, t_4 and t_5 respectively. Hence, the average MCDS node size is 2.4. Notice that the absence of link $2 - 1$ in the graph at time instant t_3 forced us to choose the MCDS with nodes $\{2, 4, 7\}$ at t_3; once this link appears at time instants t_4 and t_5, by adopting the ORA strategy – we could reduce the number of nodes in the MCDS from three to two; whereas, by adopting the LORA strategy, we end up continuing to stay with a MCDS comprising of three nodes. Updating the MCDS for every time instant helps to reduce the number of constituent CDS nodes; however, with a significant control overhead. Using a CDS with a larger number of constituent nodes leads to redundant retransmissions in the case of flooding using the CDS. This illustrates the difference and trade off between the ORA and LORA strategies.

4. REVIEW OF MOBILITY MODELS

All the three mobility models assume the network is confined within fixed boundary conditions. The Random Waypoint mobility model assumes that the nodes can move anywhere within a network region. The City Section and the Manhattan mobility models assume the network to be divided into grids: square blocks of identical block length. The network is thus basically composed of a number of horizontal and vertical streets. Each street has two lanes, one for each direction (north and south direction for vertical streets, east and west direction for horizontal streets). A node is allowed to move only along the grids of horizontal and vertical streets.

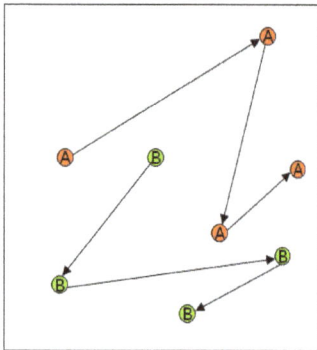

Figure 11: Movement under Random Waypoint Model

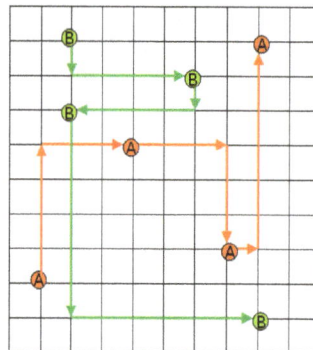

Figure 12: Movement under City Section Model

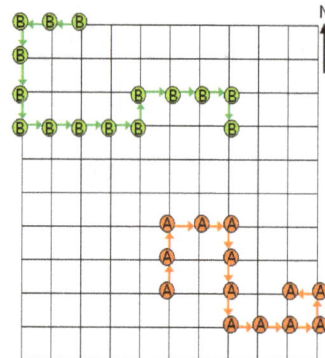

Figure 13: Movement under Manhattan Mobility Model

4.1. Random Waypoint Mobility Model

Initially, the nodes are assumed to be placed at random locations in the network. The movement of each node is independent of the other nodes in the network. The mobility of a particular node is described as follows: The node chooses a random target location to move. The velocity with which the node moves to this chosen location is uniformly and randomly selected from the interval $[v_{min},...,v_{max}]$. The node moves in a straight line (in a particular direction) to the chosen location with the chosen velocity. After reaching the target location, the node may stop there for a certain time called the pause time. The node then continues to choose another target location and moves to that location with a new velocity chosen again from the interval $[v_{min},...,v_{max}]$. The selection of each target location and a velocity to move to that location is independent of the current node location and the velocity with which the node reached that location. In Figure 11, we observe that nodes A and B move independent of each other, in random directions with randomly chosen velocities.

4.2. City Section Mobility Model

Initially, the nodes are assumed to be randomly placed in the street intersections. Each street (i.e., one side of a square block) is assumed to have a particular speed limit. Based on this speed limit and the block length, one can determine the time it would take move in the street. Each node placed at a particular street intersection chooses a random target street intersection to move. The node then moves to the chosen street intersection on a path that will incur the least amount of travel time. If two or more paths incur the least amount of travel time, the tie is broken arbitrarily. After reaching the targeted street intersection, the node may stay there for a pause time and then again choose a random target street intersection to move. The node then moves towards the new chosen street intersection on the path that will incur the least amount of travel time. This procedure is repeated independently by each node. In Figure 12, the movement of two nodes A and B according to the City Section mobility model has been illustrated.

4.3. Manhattan Mobility Model

Initially, the nodes are assumed to be randomly placed in the street intersections. The movement of a node is decided one street at a time. To start with, each node has equal chance (i.e., probability) of choosing any of the streets leading from its initial location. In Figure 13, to start with, node A has 25% chance to move in each of the four possible directions (east, west, north or south), where as node B can move only either to the west, east or south with a 1/3 chance for each direction. After a node begins to move in the chosen direction and reaches the next street intersection, the subsequent street in which the node will move is chosen probabilistically. If a node can continue to move in the same direction or can also change directions, then the node has 50% chance of continuing in the same direction, 25% chance of turning to the east/north and 25% chance of turning to the west/south, depending on the direction of the previous movement. If a node has only two options, then the node has an equal (50%) chance of exploring either of the two options. For example, in Figure 13, once node A reaches the rightmost boundary of the network, the node can either move to the north or to the south, each with a probability of 0.5 and the node chooses the north direction. After moving to the street intersection in the north, node A can either continue to move northwards or turn left and move eastwards, each with a probability of 0.5. If a node has only one option to move (this occurs when the node reaches any of the four corners of the network), then the node has no other choice except to explore that option. For example, in Figure 13, we observe node B that was traveling westward, reaches the street intersection, which is the corner of the network. The only option for node B is then to turn to the left and proceed southwards.

5. SIMULATIONS

Simulations have been conducted in a discrete-event simulator implemented by the author in Java. Network dimensions are 1000m x 1000m. For the Random Waypoint mobility model, we assume the nodes can move anywhere within the network. For the City Section and Manhattan mobility models, we assume the network is divided into grids: square blocks of length (side) 100m. The network is thus basically composed of a number of horizontal and vertical streets. Each street has two lanes, one for each direction (north and south direction for vertical streets, east and west direction for horizontal streets). A node is allowed to move only along the grids of horizontal and vertical streets. The wireless transmission range of a node is 250m. The network density is varied by performing the simulations with 50 (low density), 100 (moderate density) and 150 (high density) nodes. The node velocity values used for each of the three mobility models are 2.5 m/s (about 5 miles per hour), 12.5 m/s (about 30 miles per hour) and 25 m/s (about 60 miles per hour), representing scenarios of low, moderate and high node mobility respectively. For the Random Waypoint mobility model, we assume $v_{min} = v_{max}$.

We obtain a centralized view of the network topology by generating mobility trace files for 1000 seconds under each of the three mobility models. The network topology is sampled for every 0.25 seconds to generate the static graphs and the mobile graph. Two nodes a and b are assumed to have a bi-directional link at time t, if the Euclidean distance between them at time t (derived using the locations of the nodes from the mobility trace file) is less than or equal to the wireless transmission range of the nodes. Each data point in Figures 14 through 19 and in Tables 1 to 6 is an average computed over 5 mobility trace files and 20 randomly selected s-d pairs from each of the mobility trace files. The starting time of each s-d session is uniformly distributed between 1 to 20 seconds.

The following performance metrics are evaluated:
* *Percentage Network Connectivity*: The percentage network connectivity indicates the probability of finding an s-d path between any source s and destination d in networks for a given density and a mobility model. Measured over all the s-d sessions of a simulation run, this metric is the ratio of the number of static graphs in which there is an s-d path to the total number of static graphs in the mobile graph.
* *Average Route Lifetime*: The average route lifetime is the average of the lifetime of all the static paths of an s-d session, averaged over all the s-d sessions.
* *Average Hop Count*: The average hop count is the time averaged hop count of a mobile path for an s-d session, averaged over all the s-d sessions. The time averaged hop count for an s-

d session is measured as the sum of the products of the number of hops per static *s-d* path and the lifetime of the static *s-d* path divided by the number of static graphs in which there existed a static *s-d* path. For example, if a mobile path spanning over 10 static graphs comprises of a 2-hop static path p_1, a 3-hop static path p_2, and a 2-hop static path p_3, with each existing for 2, 3 and 5 seconds respectively, then the time-averaged hop count of the mobile path would be (2*2 + 3*3 + 2*5) / 10 = 2.3.

- *CDS Node Size*: This is a time-averaged value of the number of nodes included in the sequence of minimum connected dominating sets used over the entire duration of the simulation.

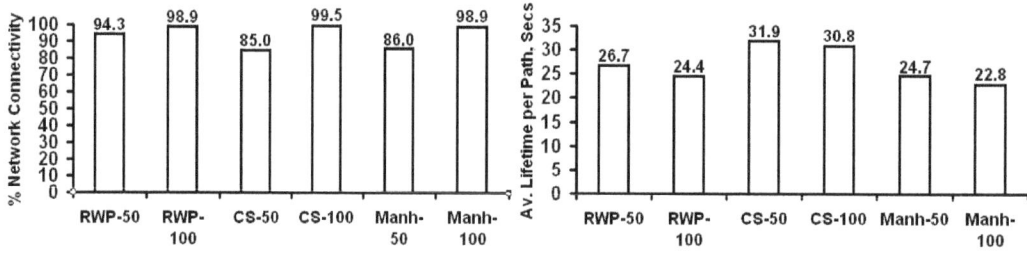

Figure 14: % Connectivity (vel = 2.5 m/s)   **Figure 15:** Lifetime per *s-d* Path (vel = 2.5 m/s)

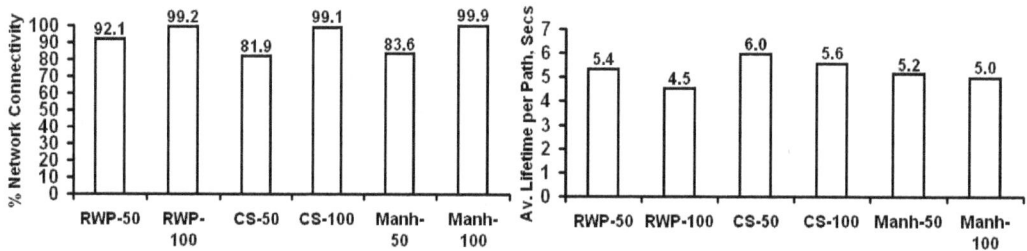

Figure 16: % Connectivity (vel = 12.5 m/s)   **Figure 17:** Lifetime per *s-d* Path (vel = 12.5 m/s)

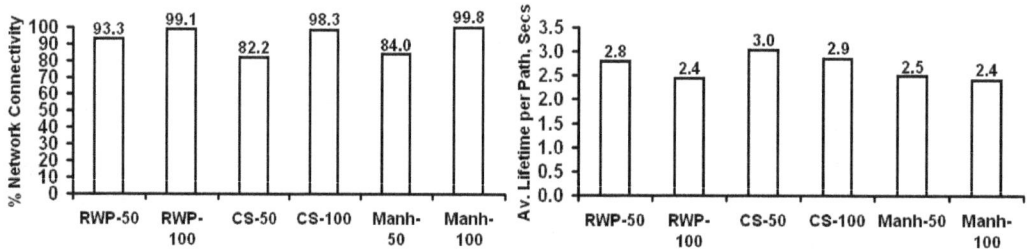

Figure 18: % Connectivity (vel = 25 m/s)   **Figure 19:** Lifetime per *s-d* Path (vel = 25 m/s)

5.1. Network Connectivity

The percentage network connectivity (refer Figures 14, 16 and 18) is not dependent on the routing strategy (ORA or LORA) and is dependent only on the mobility model, the level of node mobility and network density. It is quite natural to observe that for a given mobility model and level of node mobility, the percentage network connectivity increases with increase in network density. In low density networks (50 nodes), the Random Waypoint model provided the largest network connectivity for a given level of node mobility; the City Section and Manhattan models yielded a relatively lower network connectivity, differing as large as by 11% . This can be attributed to the constrained motion of the nodes only along the streets of the network. On the other hand, as we increase the network density (100 node scenarios), the City Section model and/or the Manhattan model yielded network connectivity equal or larger than that incurred with the Random Waypoint model. As more nodes are added to the streets, the probability of finding source-destination routes at any point of time increases significantly. It is also interesting to observe that for a given network density, the network connectivity provided by each of the three mobility models almost remained the same for different values of node velocity. Hence, network connectivity is mainly influenced by the number of nodes in the network and their initial random distribution. The randomness associated with the mobility models ensure that node velocity is not a significant factor influencing network connectivity.

5.2. Route Lifetime

The average route lifetime (Figures 15, 17 and 19) is measured only for routes discovered under the LORA strategy as routes are determined for every static graph under the ORA strategy. With LORA, a route is used as long as it exists. The average route lifetime of minimum hop routes is mainly influenced by node velocity and to a lesser extent by the mobility model and network density, in this order. For a given node velocity and network density, minimum hop routes determined under the City Section model had the largest lifetime and those determined under the Manhattan model had the smallest lifetime except the scenario of 100 nodes with 12.5 m/s velocity, wherein the Random Waypoint model yielded routes with the lowest average lifetime. For a given node velocity, the difference in the average lifetime of routes between the City Section model and the other two mobility models increase with increase in network density. The City Section model yielded a route lifetime that is 8-20% and 17-26% more than that discovered under the Random Waypoint model in low and high density networks respectively. Compared to the Manhattan model, the City Section model yielded routes that have 15-30% and 12-35% larger lifetime in low and high density networks respectively. For a given mobility model, the route lifetime seem to decrease proportionately with increase in node velocity. As we increase the node velocity from 2.5 m/s to 25 m/s, the average lifetime of minimum hop routes determined under a particular mobility model approximately reduced to $1/10^{th}$ of their value at low node velocity.

5.3. Hop Count of Minimum Hop Routes

For each mobility model, node velocity and network density, we observe that minimum hop routes discovered under the LORA strategy has a larger hop count than those discovered under the ORA strategy. But, the increase in the hop count is not substantial and is within 12%. This indicates that if the on-demand MANET routing protocols based on the LORA strategy are designed meticulously with minimum hop routing as the primary routing principle, they could discover routes that have at most 12% larger hop count than those discovered by the ORA-based proactive routing protocols. Among the three mobility models, the maximum increase in the hop count under the LORA strategy vis-à-vis the ORA strategy is observed with the Random Waypoint model and the lowest increase in the hop count is observed with the Manhattan model. However, with regards to the absolute values of the hop count, the minimum hop routes determined under the Random Waypoint model have the smallest hop count and those determined under the Manhattan model have the largest hop count.

For a given mobility model, the hop count of the minimum hop routes determined for a particular network density does not seem to be much influenced with different levels of node mobility. For a given mobility model and node velocity, we also observe that under both the ORA and LORA strategies, the average hop count of minimum hop routes decreases with increase in network density. This can be attributed to the reasoning that with a larger number of nodes in the network, there is a larger probability of finding an *s-d* path involving only fewer nodes that lie on the path from the source to the destination. The decrease in the hop count of minimum hop routes with increase in network density is very much appreciable for the Manhattan model compared to the other two mobility models.

Another interesting observation is that for a given network density, the percentage increase in the average hop count per minimum hop *s-d* path decreases with increase in node mobility. This can be attributed to significant decrease in the lifetime of the *s-d* routes with increase in node mobility. At higher node mobility, the sub-optimal routes do not exist for a longer time and the sequence of routes determined under the LORA strategy starts getting closer to the sequence of routes determined under the ORA strategy. This effect is more predominant in the case of MCDS as the lifetime per CDS under the LORA strategy is significantly smaller than the lifetime per *s-d* path.

Table 1: Average Hop Count per *s-d* Path under Random Waypoint Mobility Model

Node Velocity	50 Node Network			100 Node Network			150 Node Network		
	ORA	LORA	Percent Increase	ORA	LORA	Percent Increase	ORA	LORA	Percent Increase
2.5 m/s	2.36	2.63	11.51%	2.27	2.50	10.40%	2.21	2.43	9.95%
12.5 m/s	2.40	2.65	10.14%	2.36	2.58	9.46%	2.25	2.46	9.33%
25 m/s	2.40	2.63	9.44%	2.31	2.52	9.31%	2.24	2.44	8.93%

Table 2: Average Hop Count per *s-d* Path under City Section Mobility Model

Node Velocity	50 Node Network			100 Node Network			150 Node Network		
	ORA	LORA	Percent Increase	ORA	LORA	Percent Increase	ORA	LORA	Percent Increase
2.5 m/s	2.66	2.86	7.51%	2.45	2.71	10.40%	2.24	2.50	11.60%
12.5 m/s	2.85	3.07	7.66%	2.70	2.93	8.68%	2.55	2.80	9.80%
25 m/s	2.83	3.04	7.39%	2.60	2.82	8.47%	2.37	2.60	9.70%

Table 3: Average Hop Count per *s-d* Path under Manhattan Mobility Model

Node Velocity	50 Node Network			100 Node Network			150 Node Network		
	ORA	LORA	Percent Increase	ORA	LORA	Percent Increase	ORA	LORA	Percent Increase
2.5 m/s	3.31	3.60	8.81%	3.08	3.34	8.38%	2.75	3.03	10.12%
12.5 m/s	3.37	3.60	6.90%	3.00	3.26	8.60%	2.67	2.93	9.82%
25 m/s	3.51	3.74	6.60%	3.03	3.27	7.94%	2.52	2.76	9.35%

5.4. Node Size per Minimum Connected Dominating Sete

For each mobility model, the average node size per MCDS determined under the LORA strategy is slightly higher than that determined under the ORA strategy. But, the increase is very minimal and is only within 6%. This implies that the number of retransmissions incurred by adopting the sequence of MCDS determined under the LORA strategy will not be substantially higher than those incurred using the sequence of MCDS determined under the ORA strategy. On the other hand, there would be a significant control overhead in updating the MCDS for every time instant. Hence, the LORA strategy could always be the preferred strategy to determine and use MCDS in MANETs.

With respect to the absolute magnitude of the MCDS Node Size under the three mobility models, we observe that the MCDS Node Size determined under the Random Waypoint model is always the smallest and the MCDS Node Size determined under the Manhattan mobility model is always the largest under the different conditions of node mobility and network density. The MCDS Node Size determined under the City Section mobility model is 16%, 20%-25% and

22%-24% larger than that determined under the Random Waypoint model in conditions of low, moderate and high network density respectively. The MCDS Node Size determined under the Manhattan mobility model is 26%-36%, 31%-34% and 30%-34% larger than that determined under the Random Waypoint model in conditions of low, moderate and high network density respectively.

Table 4: Average Node Size per MCDS under Random Waypoint Mobility Model

Node Velocity	50 Node Network			100 Node Network			150 Node Network		
	ORA	LORA	Percent Increase	ORA	LORA	Percent Increase	ORA	LORA	Percent Increase
2.5 m/s	9.80	10.12	3.27%	10.17	10.62	4.47%	10.09	10.65	5.55%
12.5 m/s	9.88	10.23	3.52%	9.93	10.24	3.12%	10.41	10.75	3.25%
25 m/s	9.54	9.94	4.19%	9.81	10.15	3.47%	10.21	10.51	2.93%

Table 5: Average Node Size per MCDS under City Section Mobility Model

Node Velocity	50 Node Network			100 Node Network			150 Node Network		
	ORA	LORA	Percent Increase	ORA	LORA	Percent Increase	ORA	LORA	Percent Increase
2.5 m/s	11.43	11.78	3.06%	12.18	12.62	3.61%	12.58	13.07	3.89%
12.5 m/s	11.37	11.79	3.69%	12.18	12.65	3.86%	12.91	13.31	3.09%
25 m/s	11.15	11.47	2.87%	12.22	12.46	1.96%	12.68	12.91	1.81%

Table 6: Average Node Size per MCDS under Manhattan Mobility Model

Node Velocity	50 Node Network			100 Node Network			150 Node Network		
	ORA	LORA	Percent Increase	ORA	LORA	Percent Increase	ORA	LORA	Percent Increase
2.5 m/s	12.42	12.89	3.78%	13.33	13.61	2.10%	13.53	13.95	3.10%
12.5 m/s	12.90	13.42	4.03%	13.34	13.76	3.15%	13.58	13.98	2.94%
25 m/s	13.01	13.32	2.38%	13.11	13.39	2.14%	13.45	13.73	2.08%

As observed in the case of minimum hop routes, for a given network density, the percentage increase in the MCDS Node Size decreases with increase in node mobility. This can be attributed to the decrease in the MCDS lifetime by factors of 4 to 5 and 8 to 9 with increase in the node velocity from 2.5 m/s to 12.5 m/s and 25 m/s respectively. For a given condition of node mobility and network density, the lifetime per MCDS is only $1/3^{rd}$ to $1/4^{th}$ of the lifetime per *s-d* path determined under similar conditions. Hence, compared to the minimum hop routes, the *CDS-Node-List* of the sequence of MCDS formed under the LORA strategy fast coincides with that of the sequence of MCDS formed under the ORA strategy.

6. CONCLUSIONS AND FUTURE WORK

Our hypothesis that there would be difference in the hop count of minimum hop routes and the node size of the minimum connected dominating sets (MCDS) discovered under the ORA and LORA strategies has been observed to be true through extensive simulations, the results of which are summarized in Tables 1 through 6. However, the difference is not significantly high and is within 6-12% for minimum hop routes and at most 6% for MCDS, depending mainly on the mobility model employed and the level of node mobility and to a lesser extent on the network density. With respect to absolute values, the Random Waypoint model yields minimum hop routes with the smallest hop count and MCDS with the smallest node size; whereas, the

Manhattan model yields minimum hop routes with the largest hop count and MCDS with the largest node size. With respect to the increase in the hop count of minimum hop routes due to the use of LORA strategy vis-à-vis the ORA strategy, we observe that the Random Waypoint model incurs the maximum increase and the Manhattan model incurs the smallest increase. The City Section model is ranked in between the two mobility models with regards to the absolute value of the hop count and the relative increase in the hop count with the LORA strategy. In the context of the MCDS, the percentage increase in the number of nodes per MCDS due to the use of LORA vis-à-vis ORA is about same for all the three mobility models. With regards to the route lifetime, the minimum hop routes determined under the City Section model are relatively more stable (i.e. have larger lifetime) compared to the other two mobility models.

Another interesting observation is that for a given network density, the percentage increase in the average hop count per minimum *s-d* path and the number of nodes per MCDS decreases with increase in node mobility. This can be attributed to the significant decrease in the lifetime of the *s-d* routes and the MCDS with increase in node mobility. This effect is more predominant in the case of MCDS as the lifetime per MCDS under the LORA strategy is significantly smaller than the lifetime per *s-d* path. Hence, compared to the minimum hop routes, the *CDS-Node-List* of the sequence of MCDS formed under the LORA strategy fast coincides with that of the sequence of MCDS formed under the ORA strategy. As future work, we will be extending this study and will examine the impact of the ORA vs. LORA strategies and the three mobility models on minimum-hop based multicast routing, minimum-link based multicast Steiner trees, as well as node-disjoint and link-disjoint multi-path routing for MANETs.

REFERENCES

[1] M. Abolhasan, T. Wysocki and E. Dutkiewicz, "A Review of Routing Protocols for Mobile Ad hoc Networks," *Ad hoc Networks*, vol. 2, no. 1, pp. 1-22, January 2004.

[2] N. Meghanathan, "Survey and Taxonomy of Unicast Routing Protocols for Mobile Ad hoc Networks," *The International Journal on Applications of Graph Theory in Wireless Ad hoc Networks and Sensor Networks*, vol. 1, no. 1, pp. 1-21, December 2009.

[3] C. Siva Ram Murthy and B. S. Manoj, "Routing Protocols for Ad Hoc Wireless Networks," Ad Hoc Wireless Networks: Architectures and Protocols, Chapter 7, Prentice Hall, June 2004.

[4] J. Broch, D. A. Maltz, D. B. Johnson, Y. C. Hu and J. Jetcheva, "A Performance of Comparison of Multi-hop Wireless Ad hoc Network Routing Protocols," *Proceedings of the 4th Annual ACM/IEEE Conference on Mobile Computing and Networking*, pp. 85 – 97, October 1998.

[5] P. Johansson, T. Larsson, N. Hedman, B. Mielczarek and M. Degermark, "Scenario-based Performance Analysis of Routing Protocols for Mobile Ad hoc Networks," *Proceedings of the 5th Annual International Conference on Mobile Computing and Networking*, pp. 195 – 206, August 1999.

[6] K. M. Alzoubi, P. J. Wan and O. Frieder, "New Distributed Algorithm for Connected Dominating Set in Wireless Ad Hoc Networks," *Proceedings of the 35th Hawaii International Conference on System Sciences*, pp. 3849-3855, 2002.

[7] F. Kuhn, T. Moscibroda and R. Wattenhofer, "Unit Disk Graph Approximation," *Proceedings of the ACM DIALM-POMC Joint Workshop on the Foundations of Mobile Computing*, pp. 17-23, Philadelphia, October 2004.

[8] Y. P. Chen and A. L. Liestman, "Approximating Minimum Size Weakly-Connected Dominating Sets for Clustering Mobile Ad Hoc Networks," *Proceedings of the ACM International Symposium on Mobile Ad hoc Networking and Computing*, Lausanne, Switzerland, June 9-11, 2002.

[9] N. Meghanathan, "On the Stability of Paths, Steiner Trees and Connected Dominating Sets in Mobile Ad hoc Networks," *Ad hoc Networks*, vol. 6, no. 5, pp. 744-769, July 2008.

[10] K. M. Alzoubi, P.-J Wan and O. Frieder, "Distributed Heuristics for Connected Dominating Set in Wireless Ad Hoc Networks," *IEEE / KICS Journal on Communication Networks*, vol. 4, no. 1, pp. 22-29, 2002.

[11] S. Butenko, X. Cheng, D.-Z. Du and P. M. Paradlos, "On the Construction of Virtual Backbone for Ad Hoc Wireless Networks," *Cooperative Control: Models, Applications and Algorithms*, pp. 43-54, Kluwer Academic Publishers, 2002.

[12] S. Butenko, X. Cheng, C. Oliviera and P. M. Paradlos, "A New Heuristic for the Minimum Connected Dominating Set Problem on Ad Hoc Wireless Networks," *Recent Developments in Cooperative Control and Optimization*, pp. 61-73, Kluwer Academic Publishers, 2004.

[13] T. Camp, J. Boleng and V. Davies, "A Survey of Mobility Models for Ad Hoc Network Research," *Wireless Communication and Mobile Computing*, vol. 2, no. 5, pp. 483-502, September 2002.

[14] A. Farago and V. R. Syrotiuk, "MERIT: A Scalable Approach for Protocol Assessment," *Mobile Networks and Applications*, vol. 8, no. 5, pp. 567 – 577, October 2003.

[15] T. H. Cormen, C. E. Leiserson, R. L. Rivest and C. Stein, "Single-Source Shortest Paths," *Introduction to Algorithms*, 2nd Edition, Chapter 24, MIT Press, 2001.

3

Mobile Agent Based Congestion Control Using AODV Routing Protocol Technique for Mobile Ad-hoc Network

Vishnu Kumar Sharma [1] and Dr. Sarita Singh Bhadauria [2]

[1]Department of CSE, JUET, Guna, Madhya Pradesh, India
vishnusharma97@gmail.com
[2] Department of Elex, MITS Gwalior, Madhya Pradesh, India
Saritamits.61@yahoo.co.in

ABSTRACT

In Mobile Ad hoc Networks (MANETs) obstruction occurs due to the packet loss and it can be successfully reduced by involving congestion control scheme which includes routing algorithm and a flow control at the network layer. In this paper, we propose to agent based congestion control technique for MANETs. In our technique, the information about network congestion is collected and distributed by mobile agents (MA) A mobile agent based congestion control AODV routing protocol is proposed to avoid congestion in ad hoc network. Some mobile agents are collected in ad-hoc network, which carry routing information and nodes congestion status. When mobile agent movements through the network, it can select a less-loaded neighbor node as its next hop and update the routing table according to the node's congestion status. With the support of mobile agents, the nodes can get the dynamic network topology in time. By simulation results, we have shown that our proposed technique attains high delivery ratio and throughput with reduced delay when compared with the different existing technique.

KEYWORDS

AODV routing protocol, Congestion control Mobile Ad hoc Networks (MANETs), Mobile Agents (MA), Total Congestion Metric (TCM), Enhanced Distributed Channel Access (EDCA), Transmission opportunity limit (TXOP).

1. INTRODUCTION

The mobile ad-hoc network is accomplished of forming a temporary network, without the require of a central administration or standard support devices available in a conventional network, thus forming an infrastructure-less network. In order to guarantee for the future, the mobile ad hoc networks establishes the networks everywhere. To avoid being an perfect candidate during rescue and emergency operations, these networks do not depend on the irrelevant hardware. These networks build, operate and maintain with the help of constituent wireless nodes. Since these nodes have only a limited transmission range, it depends on its neighboring nodes to forward packets [1].

Obstruction control is a key problem in mobile ad-hoc networks. The standard TCP congestion control mechanism is not able to handle the unique properties of a shared wireless multihop channel well. In particular the frequent changes of the network topology and the shared nature of the wireless channel pose significant challenges.

Many approaches have been projected to overcome these difficulties [3] Ad-hoc network is a wireless and with no fixed apparatus (such as base stations) distributed network which is component of mobile terminals [27], each mobile terminal is not only host computer but also router.

As power and bandwidth restrictions, ad-hoc network routing protocols should allocate routing tasks literally in the mobile nodes. At present, AODV routing protocol is often used in ad-hoc network. But its biggest failing is delay. In routing discovery and maintenance, a large number of data is transmitted through a small number of nodes is hop to lead to network congestion and bottleneck. At the same time, unwarranted data load will be exhaust nodes energy rapidly. With the increase of brownout nodes, network connectivity will be weakened and network overall survival time will be shorten subsequently. Therefore, In order to balance the network load and maintain network continuous, efficient and stable operation, it is necessary to take into account the routing nodes load and congestion in network [28]. In mobile ad hoc wireless network, mobile agent has mobility and autonomy. Therefore, it can be used to solve the ad hoc network congestion [26].

Ad-hoc networks are characterized by a need of infrastructure, and by a random and quickly varying network topology; thus the need for a robust dynamic routing protocol that can accommodate such an environment. Therefore, many routing algorithms have come into existence to satisfy the needs of communications in such networks. Recital comparison between two routing algorithms, AODV, from the immediate family and DSDV, from the proactive family. Both protocols were simulated using the ns-2 and were compared in terms of average throughput, packet loss ratio, and routing overhead, while changeable number of nodes, speed and pause time. Simulation exposed that although DSDV completely scales to small networks with low node speeds, AODV is favored due to its more efficient use of bandwidth [31].

1.2 Congestion Control in MANETs

Congestion takes place in MANETs with limited resources. In these networks, shared wireless channel and dynamic topology leads to interference and fading during packet transmission. Packet victims and bandwidth dilapidation are caused due to congestion, and thus, time and energy is wasted during its recovery. Congestion can be prevented using congestion-aware protocol through bypassing the affected links [2]. Severe throughput degradation and massive fairness problems are some of the identified congestion related problems. These problems are generated from MAC, and protocol routing and transport layers [3].

Congestion control is the main problem in ad-hoc networks. Congestion control is associated to controlling traffic incoming into a telecommunication network. To avoid congestive crumple or link capabilities of the intermediate nodes and networks and to reduce the rate of sending packets congestion control is used extensively [4]. Congestion control and dependability mechanisms are combined by TCP to perform the congestion control without explicit feedback about the congestion position and without the intermediate nodes being directly intermittent [4]. Their principles include packet conservation, additive increase and multiplicative decrease in sending rate, stable network. End system flow control, network congestion control, network based congestion avoidance, and resource allotment includes the basic techniques for congestion control [5].

Packet failure in MANETs is primarily caused due to obstruction. The packet loss can be condensed by involving congestion control over a mobility and failure adaptive routing protocol at the network layer. The congestion non-adaptive routing protocols, leads to the following difficulties:

- Long delay: The congestion control mechanisms takes much time for detecting congestion. Usage of new routes in some critical situations is advisable. In an on-demand routing protocol, the main problem is the delay stirring for route searching.

- High overhead: It takes effort in new routes for processing and communication for discovering it. It also takes effort in multipath routing for maintaining the multi-paths, though there is another protocol.

- Many packet losses: The packets may be lost when the congestion is detected. To decrease the traffic load, a congestion control solution is applied either by decreasing the sending rate at the sender, or dropping packets at the intermediate nodes or by both methods. But high packet loss rate or a small throughput occurs at the receiver [6].

1.3 Problem discovery and Proposed Protocol Overview

Congestion adaptive routing has been examined in several studies. Estimating or reviewing the level of activity in the intermediate nodes using load or delay measurement, is the common approach in all the studies mentioned. The favorable path is established based upon the collected information, which helps in avoiding the existing and developing congested nodes. The performance of routing protocols is affected by the service type of the traffic carried by the intermediate nodes. But no research has stated this so far.

Before presenting themselves as aspirant to route traffic to the destination, the MANETs do not take the status of the queues into account, for the route discovery process. Because of this, the newly arriving traffic face long delays, packet drops, and fail to be transmit ahead of the already queuing traffic.

The mobile ad hoc networks performances are subjective to the congestion problem. A routing algorithm and a flow control scheme, includes the congestion control scheme. Enhanced performance and better congestion control can be achieved only by considering the routing and the flow control together. This was not done in earlier researches [12].

AODV routing protocol [30] is a distance vector routing protocol based on demand. The main characteristic is using a serial number to identify the routing is new or old and avoid routing loop. AODV routing protocol has the advantages of each intermediate node saves a routing request and response result implicitly to adapt to dynamic link rapidly .AODV protocol has two main components: routing discovery and routing maintenance.

2. Related Work

Yao-Nan Lien et al [7] proposed a new TCP congestion control mechanism by router-assisted approach. Their proposed TCP protocol, called TCP Muzha uses the assistance provided by routers to achieve better congestion control. To use TCP Muzha, routers are required to provide some information allowing the sender to estimate more accurately the remaining capacity over the bottleneck node with respect to the path from the sender to the receiver. With this information, TCP Muzha will be able to enhance the performance of both TCP and network.

Wei Sun et al [8] have compared the general AIMD-based congestion control mechanism (GAIMD) with Equation-based congestion control mechanism (TFRC TCP-Friendly Rate Control) over a wide range of MANET scenario, in terms of throughput fairness and smoothness. Their results have shown that TFRC and GAIMD are able to maintain throughput smoothness in MANET, but at the same time, they require only a less throughput than the competing TCP flows. Also their results show that TFRC changes its sending rate more smoothly than GAIMD does, but it gets the least throughput compares with TCP and GAIMD.

Yung Yi et al [9] have developed a fair hop-by-hop congestion control algorithm with the MAC constraint being imposed in the form of a channel access time constraint, using an optimization-based framework. In the absence of delay, they have shown that their algorithm is globally stable using a Lyapunov-function-based approach. Next, in the presence of delay, they have shown that the hop-by-hop control algorithm has the property of spatial spreading. Also they have derived bounds on the "peak load" at a node, both with hop-by-hop control, as well as with end-to-end control, show that significant gains are to be had with the hop-by-hop scheme, and validate the analytical results with simulation.

Umut Akyol et al [10] have studied the problem of jointly performing scheduling and congestion control in mobile adhoc networks so that network queues remain bounded and the resulting flow rates satisfy an associated network utility maximization problem. They have defined a specific network utility maximization problem which is appropriate for mobile adhoc networks. They have described a wireless Greedy Primal Dual (wGPD) algorithm for combined congestion control and scheduling that aims to solve this problem. They have shown how the wGPD algorithm and its associated signaling can be implemented in practice with minimal disruption to existing wireless protocols.

S.Karunakaran et al [11] have presented a Cluster Based Congestion Control (CBCC) protocol that consists of scalable and distributed cluster-based mechanisms for supporting congestion control in mobile ad hoc networks. The distinctive feature of their approach is that it is based on the self-organization of the network into clusters. The clusters autonomously and proactively monitor congestion within its localized scope.

S.Venkatasubramanian et al [19] proposed the QoS architecture for Bandwidth Management and Rate Control in MANETs. The bandwidth information in the architecture can be used for QoS capable routing protocols to provide support to admission control. The traffic is balanced and the network capacity is improved as the weight value assists the routing protocol to evade routing traffic through congested area. The source nodes then perform call admission control for different priority of flows based on the bandwidth information provided by the QoS routing. In addition to this, a rate control mechanism is used to regulate best-effort traffic, whenever network congestion is detected. In this mechanism, the packet generation rate of the low-priority traffic is adjusted to incorporate the high-priority traffic.

R.Mynuddin Sulthani et al [20] proposed a joint design of reliable QoS architecture for mobile adhoc networks. In the reliable multipath routing protocol, dispersion and erasure code techniques are utilized for producing replicated fragments for each packet, to enhance reliability. Then messages with good delivery probability are identified and transmitted through the paths with high average node delivery index. While it receives an assured number of fragments, destination can recover the original packet. Next, a call admission control (CAC) scheme has been developed, in which, the calls are admitted based on the bandwidth availability of the path. Once congestion occurs, the best effort traffic is rate controlled, to free bandwidth for the real-time flows.

Lijun Chen et al [21] proposed the joint design of congestion control, routing and scheduling for ad hoc wireless networks. They formulate resource allocation in the network with fixed wireless channels or single-rate wireless devices as a utility maximization problem with schedulability and rate constraints arising from contention for the wireless channel. We also extend the dual algorithm to handle the network with time-varying channel and adaptive multi-rate devices, and surprisingly show that, despite stochastic channel variation, it solves an ideal reference system problem which has the best feasible rate region at link layer. In future, they will extend the results to networks with more general interference models and/or node mobility and further will enhance the performance gain from cross-layer design involving link layer.

Xuyang Wang et al [22] proposed a cross layer hop by hop congestion control scheme to improve TCP performance in multihop wireless networks which coordinates the congestion response across the transport, network, and transport layer protocols. The proposed scheme attempts to determine the actual cause of a packet loss and then coordinates the appropriate congestion control response among the MAC network, and transport protocols. The congestion control efforts are invoke at all intermediate and source node along the upstream paths directed from the wireless link experiencing the congestion induced packet drop.

Kazuya Nishimura et al [23] proposed a routing protocol that reduces network congestion for MANET using multi-agents. They use two kinds of agents: Routing Agents to collect information about congestion and to update the routing table at each node, and Message Agents to move using this information. In the future, they will investigate a better evaluation function and discuss the limits of its effectiveness. The evaluation function itself may change depending on the environment. Incorporating learning into the function is also an interesting issue.

Bhadauria,Sharma [24] proposed the information about network congestion is collected and distributed by mobile agents (MA). The MA measures the queue length of the various traffic classes and the channel contention and estimates the total congestion metric to find the minimum congestion level in the network. The congestion metric is applied in the routing protocol to select the minimum congested route.

P.K. Suri et al [29] proposed a bandwidth-efficient power aware routing protocol "QEPAR". The routing protocol is presented to minimize the bandwidth consumption as well as delay. QEPAR will help in increasing the throughput by decreasing the packet loss due to non availability of node having enough battery power to retransmit the data packet to next node. The proposed protocol is also helpful in finding out an optimal path without any loop

Vinay Rishiwal et al [30] proposed QoS based power aware routing protocol (Q-PAR). The selected route is energy stable and satisfies the bandwidth constraint of the application. The protocol Q-PAR is divided in to two phases. In the first route discovery phase, the bandwidth and energy constraints are built in into the DSR route discovery mechanism. In the event of an impending link failure, the second phase, a repair mechanism is invoked to search for an energy stable alternate path locally. Moreover the local repair mechanism was able to find an alternate path in most of the cases enhanced the network lifetime and delayed the repair and reconstruction of the route.

3. Agent Based Congestion Control Routing

3.1. EDCA Mechanism of 802.11e

The PCF and DCF modes have been replaced with HCF controlled channel access (HCCA), and enhanced distributed channel access (EDCA) which provides distributed access supplying service differentiation [13].

An extended version of the legacy DCF mechanism is EDCA. Access Categories (AC) or traffic priority classes like voice, video, best effort and background are defined by EDCA [14]. The access categories prioritize themselves from AC3 to AC0. In general, best effort and background traffic are maintained by AC1 and AC0 and real-time applications like voice or video transmission are maintained by AC2 and AC3 [15]. For the purpose of service differentiation, many MAC constraints vary with priority level chosen for each AC.

For the implementation of the EDCA contention algorithm the four transmission queues are applied with each AC being communicated with the others. The minimum idle delay before contention (AIFS), the Contention Windows (CWmin and CWmax), and the Transmission opportunity limit (TXOP) are the various parameters described here. The default values of each parameter are listed in Table 1.

Table 1. IEEE 802.11e EDCA MAC System Parameters

Access Category	AIFSN	CWmin	CWmax	Queue length	Max. retry limit
AC3	2	7	15	25	8
AC2	2	15	31	25	8
AC1	3	31	1023	25	4
AC0	7	31	1023	25	4

In the MAC layer, voice traffic is conveyed through AC3 and the video traffic is conveyed through AC2 in accordance with 802.11e EDCA standard. The AC class differentiation in EDCA is very much useful in providing services to the traffic. Superior servicing is done for high-priority traffic and not much importance is given for low-priority traffic. The contention parameters of EDCA are not able to adapt to the network conditions, in spite of the delay sensitivity of real-time traffic taken into account. This leads to limitations in the QOS improvement [16].

ACs pause for diverse values of Arbitration Interframe Space (AIFS) and AIFSi is computed by,

$$AIFSi = SIFS + AIFSNi \times SlotTime$$

where $AIFSi$ is a positive integer which is greater than one, $AIFSNi$ is the AC-specific AIFS number; SIFS and Slot Time are dependent on physical layer [14]. If the values of the subsequent parameters are small, the channel access delay will become less for the AC which leads the higher priority to approach the medium.

When a particular QoS station (QSTA) has the concession to begin transmissions, then the TXOP is expressed as the time interval in IEEE 802.11e. The initiation of the TXOP and the multiple frame transmission within an EDCA TXOP are the nodes approved by TXOP. The former occurs only when the EDCA rules allow entry to the medium. And the later occurs when an EDCA Function (EDCAF) holds the concession to contact the medium after completing a frame exchange sequence. The period of TXOP values are herewith in the EDCA parameter engraved in beacon frames. A STA is allowed to transmit multiple MAC protocol data units (MPDUs) from the same AC with a SIFS time interval between an ACK and the succeeding frame transmission. A single MPDU may be forwarded for each TXOP if the TXOP limits the value of 0 [17].

3.2 Mobile Agent (MA)

Each node has a routing table that stores k fresh routing information records from itself to every node $S : [S, \{(Tci, NHi, ANi, NPi) \cdots (Tcm, NHm, ANm, NPm)\}]$, where $Tc1 > Tc2 > \cdots > Tm$. We call m the number of entries. For each $i(1 \leq i \leq m)$, Tci is a time of visiting the adjacent node ANi, NHi is the number of hops and NPi is the number of MAs on ANi. When MA with the history (S, Tc, NH, AN, NP) visits a node N, the routing information on that node

$[S, \{(Tci, NHi, ANi, NPi) \cdots (Tcm, NHm, ANm, NPm)\}]$ is updated to

$[S; \{(Tc, NH; AN, NP), (Tci, NHi, ANi, NPi, NPi) \cdots$

$(Tcm - 1, NHm - 1, ANm - 1, NPm - 1)\}]$

3.3 Queue Length Estimation

Our goal is to acquire macroscopic network statistics using a heuristic approach. We compute the traffic rate as follows: Let the value L_o represent the offered load at the queue of node i and it is defined as

$$L_{oi} = \frac{AR_i}{SR_i} \tag{1}$$

where AR_i is the aggregate arrival rate of the packets produced and forwarded at node i while SR_i is the service rate at node i, i.e., $SR_i = 1/T$ where T is the computed exponentially weighted moving average of the packets' waiting time at the head of the service queue. The distribution of the queue length $PR(Q_1)$ (essentially this is the probability that there are Q_l packets in the queue) at the node is computed as

$$PR(Q_1) = (1 - L_{oi})L_{oi}^1 \tag{2}$$

For N distinct queues, the joint distribution is the product

$$PR(Q_{11}, Q_{12} \cdots Q_{1N}) = \prod_{i=1}^{N}(1 - L_{oi})L_{oi}^{li} \tag{3}$$

3.4 Channel Contention Estimation

IEEE 802.11 MAC with the distributed coordination function (DCF). It has the packet sequence as request-to-send (RTS), clear-to-send (CTS), data and acknowledgement (ACK). The amount of time between the receipt of one packet and the transmission of the next is called a short inter frame space (SIFS). Then the channel occupation due to MAC contention will be

$$C_{OCC} = t_{RTS} + t_{CTS} + 3t_{SIFS} + t_{acc} \tag{4}$$

Where t_{RTS} and t_{CTS} are the time consumed on RTS and CTS, respectively and t_{SIFS} is the $SIFS$ period. t_{acc} is the time taken due to access contention.

The channel occupation is mainly dependent upon the medium access contention, and the number of packet collisions. That is, C_{occ} is strongly related to the congestion around a given node.

C_{occ} can become relatively large if congestion is incurred and not controlled, and it can dramatically decrease the capacity of a congested link.

3.5 Total Congestion Metric

The Total Congestion Metric (TCM) can be estimated from the obtained queue length and the channel contention.

$$TCM = PR(Q_1) + C_{occ} \tag{5}$$

3.6 Agent Based Congestion Control Routing

The agent based congestion routing Architecture can be explained from the following figure:

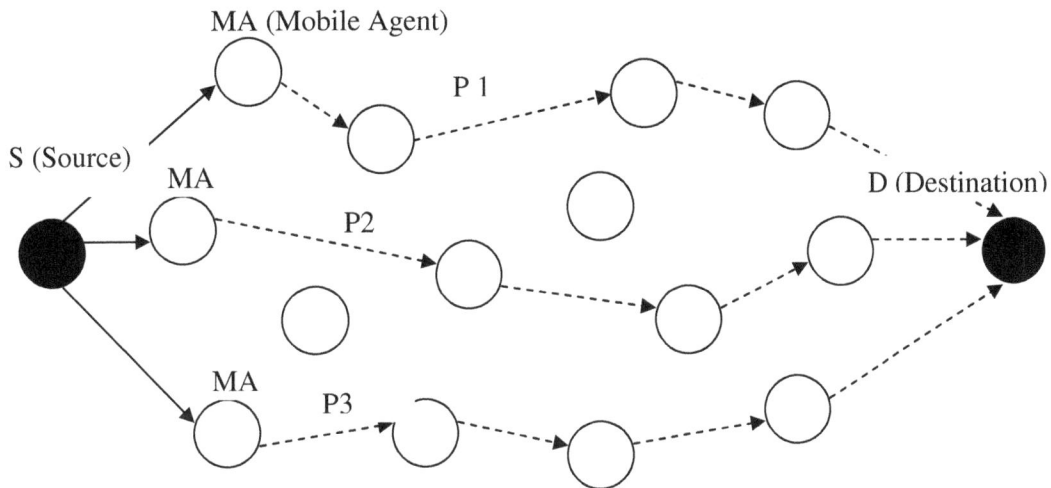

Figure 1 Agent Based Congestion Routing Architecture

Step 1: The source S checks the number of available one hop neighbors and clones the Mobile Agent (MA) to that neighbors.

Step 2: The Mobile Agent selects the shortest path of the route to move towards the destination D as given in the figure 1 such as P1, P2 and P3.

Step 3: The MA1 moves towards the destination D in a hop-by-hop manner in the path P1 and MA2 in P2 and MA3 in P3 respectively.

Step 4: Then the MA1 calculates the TCM1 of that path P1 and similarly MA2 calculates the TCM2 of P2 and MA3 calculates the TCM3 of P3.

Step 5: Now the destination D sends the total congestion metrics TCM1, TCM2 and TCM3 of the paths P1, P2 and P3 respectively to the source.

Step 6: Now the source selects path using min (TCM1, TCM2, and TCM3) and sends the data through the corresponding path which has the minimum congestion.

4. Simulation Results

4.1 Simulation Model and Parameters

We use NS2 [18] to simulate our proposed technique. In the simulation, the channel capacity of mobile hosts is set to the same value: 11Mbps. In the simulation, mobile nodes move in a 1000 meter x 1000 meter region for 50 seconds simulation time. Initial locations and movements of the nodes are obtained using the random waypoint (RWP) model of NS2. It is assumed that each node moves independently with the same average speed. All nodes have the same transmission range of 250 meters. The node speed is 5 m/s. and pause time is 5 seconds. In the simulation, for class1 traffic video is used and for class2 and Class3, CBR and FTP are used respectively.

The simulation settings and parameters are summarized in table 2.

Table 2. Simulation Settings

No. of Nodes	10, 20, 50 and 100
Area Size	1000 X 1000
Mac	802.11e
Radio Range	250m
Simulation Time	50 sec
Routing Protocol	AODV
Traffic Source	CBR and Video
Video Trace	JurassikH263-256k
Packet Size	512 byte
Mobility Model	Random Way Point
Speed	5m/s
Pause time	5 sec
MSDU	2132
Varying Rates	250kb,500kb,.....1000Kb
Varying No. of Flows	2,4,6,8 and 10

4.2 Performance Metrics

The performance of proposed Agent Based Congestion Control (ABCC) technique with the Hop by Hop algorithm [9], Cluster Based Congestion Control (CBCC) [11], Congestion-Aware Routing Protocol for Mobile Ad Hoc Networks (CARM) [2], Ad-hoc On-Demand Distance Vector (AODV) Routing [33], Congestion Aware Routing Protocol (CARP) [1], QoS Architecture for Resource Provisioning and Rate Control (QARP-RC) [34] has been compared. The performance is evaluated mainly, according to the following metrics.

Packet Delivery Fraction: It is the ratio of the number of packets received successfully and the total number of packets sent.

Throughput: It is the number of packets received successfully.

Average end-to-end delay: The end-to-end-delay is averaged over all surviving data packets from the sources to the destinations.

4.3 Results

A. Effect of Varying Rates

In the initial experiment, we measure the performance of the proposed technique by varying the rate as 250, 500, 750 and 1000Kb.

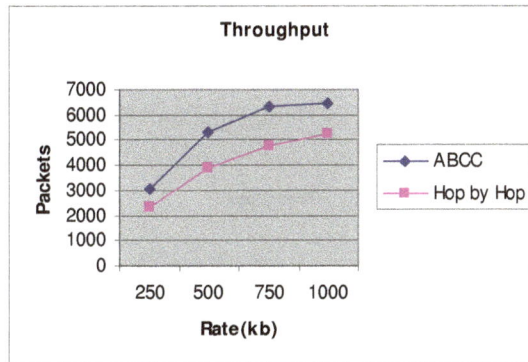

Fig. 2 Rate Vs Throughput for 10 nodes

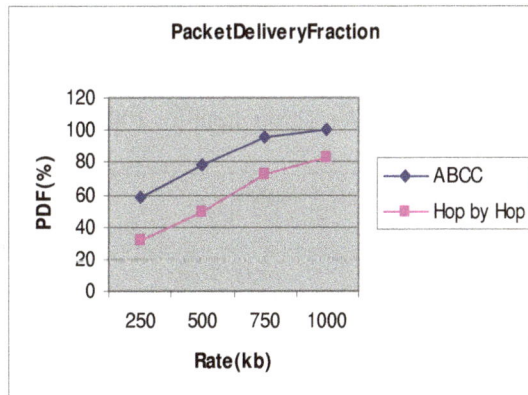

Fig. 3 Rate Vs Packet Delivery Fraction for 20 nodes

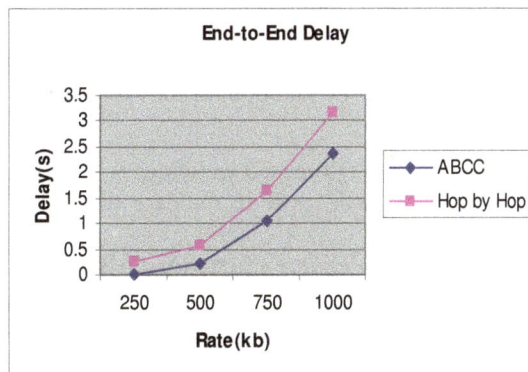

Fig. 4 Rate Vs End-to-End Delay for 50 nodes

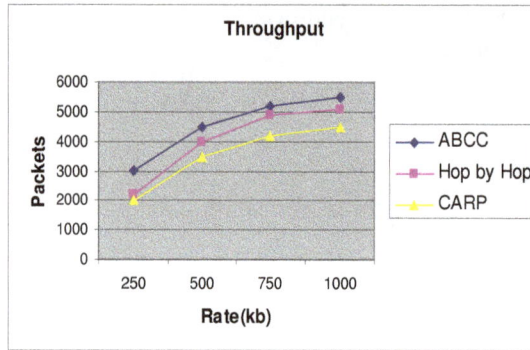

Figure 5 Rate Vs Throughput for 100 nodes

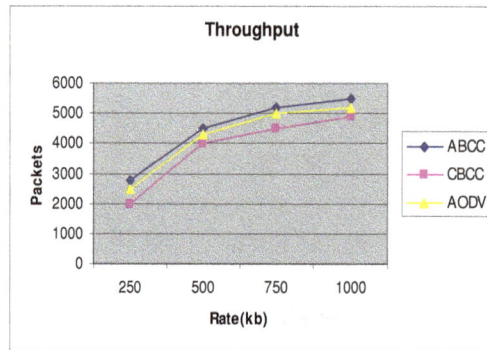

Figure 6 Rate Vs Throughput for 100 nodes

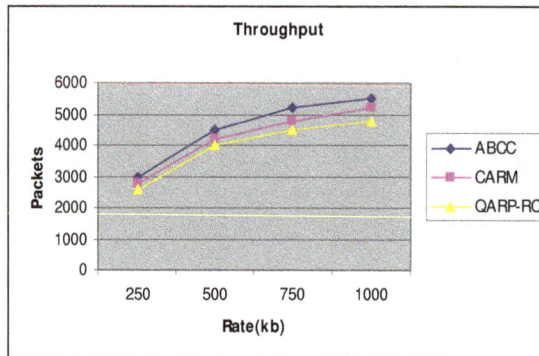

Figure 7 Rate Vs Throughput for 100 nodes

Fig 2, 5, 6, 7 gives the throughput of the proposed technique when the rate is increased. As we can see from the figure, the throughput is more in the case of ABCC when compared to the existing technique.

From Fig 3, we can see that the packet delivery fraction for ABCC is more, when compared to the Hop by Hop algorithm.

From Fig 4, we can see that the average end-to-end delay of the proposed ABCC technique is less when compared to the Hop by Hop algorithm.

B. Effect of varying Flows

In the next experiment, we compare our proposed technique by varying the number of flows as 2, 4, 6, 8 and 10.

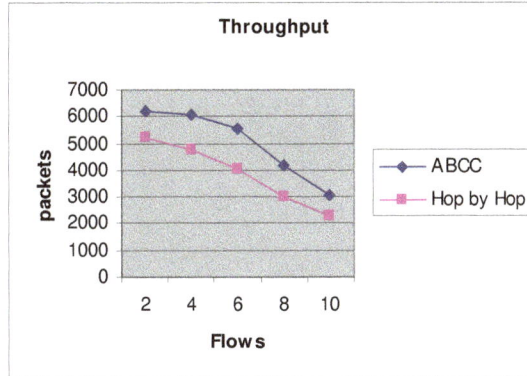

Fig. 8 Flows Vs Throughput for 10 nodes

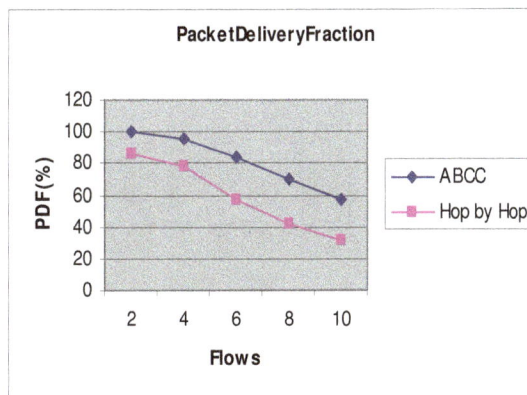

Fig. 9 Flows Vs Packet Delivery Fraction for 20 nodes

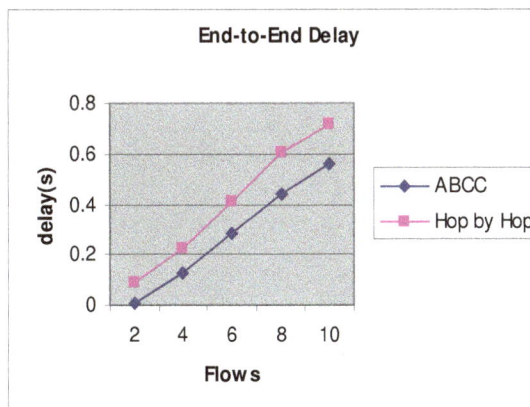

Fig. 10 Flows Vs End-to-End Delay for 50 nodes

100 nodes

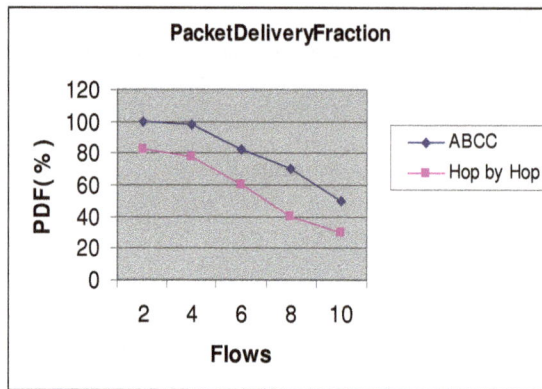

Figure 11 Flows Vs Packet Delivery Fraction for 100 nodes

From Figure 11, it is seen that the packet delivery fraction for ABCC is more, when compared to the Hop by Hop algorithm.

100 nodes

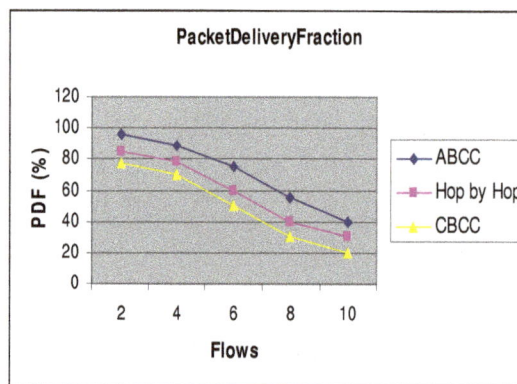

Figure 12 Flows Vs Packet Delivery Fraction for 100 nodes

From Figure 12, it is seen that the packet delivery fraction for ABCC is more, when compared to the Hop by Hop algorithm and CBCC.

100 nodes

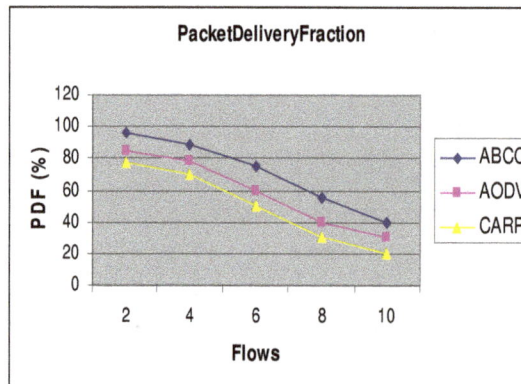

Figure 13 Flows Vs Packet Delivery Fraction for 100 nodes

Fig 8 gives the throughput of the proposed technique when the flow is increased. As we can see from the figure, the throughput is more in the case of ABCC when compared to the Hop by Hop algorithm.

From Fig 9, we can see that the packet delivery fraction for ABCC is more, when compared to the Hop by Hop algorithm.

From Fig 10, we can see that the average end-to-end delay of the proposed ABCC technique is less when compared to the Hop by Hop algorithm.

From Figure 13, it is seen that the packet delivery fraction for ABCC is more, when compared to the AODV algorithm and CARP algorithm.

5. Conclusion

In this paper, we have developed of an agent based congestion control technique. In our technique, the information about network congestion is collected and distributed by mobile agents (MA). A mobile agent starts from every node and moves to an adjacent node at every time. A node visited next is selected at the equivalent probability. The MA brings its own history of movement and updates the routing table of the node it is visiting. The MA updates the routing table of the node it is visiting. In this technique, the node is classified in one of the four categories depending on whether the traffic belongs to background, best effort, video or voice AC respectively. Then MA estimates the queue length of the various traffic classes and the channel contention of each path. Then this total congestion metric is applied to the routing protocol to select the minimum congested route in the network. , a mobile agent based congestion control AODV routing protocol reduces the end-to-end delay and the number of route discovery requests, balances the traffic load. By simulation results, we have shown that our proposed technique attains high delivery ratio and throughput with reduced delay when compared with the existing technique.

Reference

[1] Baboo, S.S., Narasimhan, B.: A Hop-by-Hop Congestion-Aware Routing Protocol for Heterogeneous Mobile Ad-hoc Networks. International Journal of Computer Science and Information Security. (2009).

[2] Chen, X., Jones, H.M., Jayalath, A.D.S. : Congestion-Aware Routing Protocol for Mobile Ad Hoc Networks. IEEE 66th Conference in Vehicular Technology. (2007).

[3] Lochert, C., Scheuermann, B., Mauve, M.: A Survey on Congestion Control for Mobile Ad-Hoc Networks. Wireless Communications and Mobile Computing, InterScience. (2007)

[4] http://en.wikipedia.org/wiki/Congestion_control

[5] http://www.linktionary.com/c/congestion.html

[6] Tran, D.A., Raghavendra, H.: Congestion Adaptive Routing in Mobile Ad Hoc Networks. IEEE Transactions on Parallel and Distributed Systems. (2006)

[7] Lien, Y.N., Hsiao, H.C.: A New TCP Congestion Control Mechanism over Wireless Ad Hoc Networks by Router-Assisted Approach. 27th IEEE International Conference on Distributed Computing Systems Workshops. (2007)

[8] Sun, W., Wen, T., Guo, Q.: A Performance Comparison of Equation-Based and GAIMD Congestion Control in Mobile Ad Hoc Networks. International Conference on Computer Science and Software Engineering. (2008)

[9] Yi, Y., Shakkottai, S.: Hop-by-Hop Congestion Control Over a Wireless Multi-Hop Network. IEEE/ACM Transactions on Networking. (2007)

[10] Akyol, U., Andrews, M., Gupta, P., Hobby, J., Saniee, I., Stolyar, A.: Joint Scheduling and Congestion Control in Mobile Ad-Hoc Networks. Proceedings of IEEE INFOCOM. (2008)

[11] Karunakaran, S., Thangaraj, P.: A Cluster Based Congestion Control Protocol for Mobile Adhoc Networks. International Journal of Information Technology and Knowledge Management. Vol. 2, No. 2, pp. 471-474.(2010).

[12] Malika, B., Mustapha, L., Abdelaziz, M., Nordine, T., Mehammed, D., Rachida, A.: Intelligent Routing and Flow Control In MANETs. Journal of Computing and Information Technology. doi: 10.2498/cit.1001470

[13] Li, J., Li, Z., Mohapatra, P.: APHD: End-to-End Delay Assurance in 802.11e Based MANETs. Mobile and Ubiquitous Systems – Workshops, 3rd Annual International Conference. pp. 1-8. (2006)

[14] Lee, J.F., Liao, W., Chen, M.C. : A Differentiated Service Model for Enhanced Distributed Channel Access (EDCA) of IEEE 802.11e WLANs. in Proc. IEEE Globecom. (2005)

[15] Ksentini, A., Naimi, M., Gueroui, A.: Toward an Improvement of H.264 Video Transmission over IEEE 802.11e through a Cross-Layer Architecture. IEEE Communications Magazine. (2006)

[16] Wu, Y-J., Chiu, J-H., Sheu, T-L. : A Modified EDCA with Dynamic Contention Control for Real-Time Traffic in Multi-hop Ad Hoc Networks. Journal of Information Science and Engineering. vol. 24, pp.1065-1079. (2008).

[17] Flaithearta, P.O, Melvin, H.:802.11e EDCA Parameter Optimization Based on Synchronized Time. MESAQIN. (2009)

[18] Network Simulator, http://www.isi.edu/nsnam/ns

[19] Venkatasubramanian, S., Gopalan, N.P.: A Quality of service architecture for resource provisioning and rate control in mobile ad hoc networks. International Journal of Ad hoc, Sensor & Ubiquitous Computing (IJASUC). Vol.1, No.3.(2010).

[20] Sulthani, R.M., Rao, D.S.: Design of an Efficient QoS Architecture (DEQA) for Mobile Ad hoc Networks. ICGST-CNIR Journal. Vol. 8, Issue 2. (2009).

[21] Chen, L., Lowy, S.H., Chiangz, M., Doyley, J.C.: Cross-layer Congestion Control, Routing and Scheduling Design in Ad Hoc Wireless Networks. Proc., IEEE, 25th international conference on computer communication, INFOCOM. pp 1 – 13. (2007)

[22] Wang, X., Perkins, D.: Cross-layer Hop-by-hop Congestion Control in Mobile Ad Hoc Networks. Proc., IEEE, wireless communication and networking conference, WCNC. pp 2456 – 2461. (2008)

[23]   Nishimura, K., Takahashi, K.: A Multi-Agent Routing Protocol with Congestion Control for MANET. Proceedings 21st European Conference on Modeling and Simulation (ECMS). (2007)

[24] Bhadauria SS,Sharma V.,Framework and Implimentation of an Agent Based Congestion Control Technique for Mobile Ad-hoc Network ,ICAC3 2011, CCIS 125, pp. 318–327, 2011.

[25] Sharma V, Bhadauria SS," Agent Based Congestion Control Routing for Mobile Ad-hoc Network"Wimon-2011,CCIS,Volume 197, pp.324-333.

[26] Hong Li, Chu Dan, Wang Min, Li Shurong,"Mobile Agent Based Congestion Control AODV Routing Protocol" Wireless Communications, Networking and Mobile Computing, 2008. WiCOM '08. 4th International Conference, Dalian,12-14 Oct. 2008.

[27] RAMANATHAN R, REDI J. "A Brief Overview of Ad Hoc Networks Challenges and Directions" IEEE Communications Magazine 2002, 40(5),20—22.

[28] HASSANEIN H, ZHOU A. Routing with Load Balancing in Wireless Ad- Hoc Networks[C]. Proceedings of the 4th ACM International Workshop on Modelling, Analysis , and Simulation of Wireless and Mobile.   Systems, Rome,2001:89-96.

[29] Thomas Kunz," Energy-Efficient MANET Routing: Ideal vs. Realistic Performance", International conference on wireless communication and mobile computing (IWCMC), pp 786 – 793, 2008.

[30] Vinay Rishiwal, Shekhar Verma and S. K. Bajpai, "QoS Based Power Aware Routing in MANETs," International Journal of Computer Theory and Engineering (IJCTE), vol. 1, no. 1, pp. 47-54, 2009.

[31] PERKINS C E, ROYER E M, DAD S R. Ad Hoc On-demand Distance   Vector(AODV) Routing[C]. IEEE Workshop on Mobile Computing   Systems and Applications, 1999:90-100.

[32] Ali El-Haj-Mahmoud, Rima Khalaf, Ayman Kayssi, "PERFORMANCE COMPARISON OF THE AODV AND DSDV ROUTING PROTOCOLS IN MOBILE AD HOC NETWORKS" Communication Systems and Networks, September 9 – 12,Spain, 2002

[33] C. E. Perkins and E. M. Royer, "The Ad Hoc On-Demand Distance-Vector Protocol (AODV)," In Ad Hoc Networking, C. E. Perkins (Ed.), Addison-Wesley, pp. 173–219, 2001.

[34] Venkatasubramanian, S., Gopalan, "N.P.: A Quality of service architecture for   resource provisioning and rate control in mobile ad hoc networks", International Journal of Ad hoc, Sensor & Ubiquitous Computing (IJASUC),Vol. 1, No.3, pp. 106-120, 2010.

4

SAMPCAN:
A NOVEL CACHING TECHNIQUE FOR CLIENT-SERVER INTERACTION MODEL IN LARGE AD HOC NETWORKS USING RESAMPLING METHODS

Paramasiven Appavoo

Dept. of Computer Science & Engineering, University of Mauritius, Réduit, Mauritius
p.appavoo@uom.ac.mu

ABSTRACT

The scalability of routing protocols has been a prominent topic to increasing the lifetime of a mobile ad hoc network. One of the underlying features behind scalable protocols is the localised nature of path maintenance. However, nodes' data communications are excluded, despite being the main purpose of the routing protocols. SAMPCAN is a novel caching technique that can be applied to numerous applications of the client-server interaction model for an optimised storage and transmission of data. SAMPCAN uses a subsampling method that can be applied to two data types, mainly (1) XML document, and (2) Image. This novel method promotes the locality of nodes cooperation and interactions to fulfil requests of neighbouring nodes. In this paper, the application of SAMPCAN is discussed for the scalability and increased lifetime of mobile ad hoc networks.

KEYWORDS

Caching, routing, scalability, ad hoc network, client server

1. INTRODUCTION

1.1. Mobile computing with ad hoc network

Mobile computing has revolutionized the world of e-services by allowing people to roam and remain connected allowing them to have access to same service anywhere and anytime. This has mainly led to increased productivity and efficiency. While the driving technologies behind are mainly Mobile IP and wireless communication, miniaturization is allowing powerful devices to be portable. Moreover, according to Moore's Law, the number of active devices that can be placed over a given area of silicon doubles every 18 months. This implies that the evolution and proliferation of mobile computing is far from an end to itself.

Current research is also focusing on the seamless integration of portable devices with their environment in an ad hoc manner. This has been made possible with standard ad hoc mode of wireless communication like mainly Bluetooth (IEEE 802.15.1) and WiFi (IEEE 802.11 b/g), which are quite common with their widespread range of applications in several domains. While Bluetooth is short range and is mainly use for Personal Area Network, with WiFi support, it is possible to have wireless LAN. Heterogeneous devices from different vendors just need to support any of such standards to allow for interoperability. In addition, service discovery protocols have been developed alongside with the standards which mainly allow the deployment of services and access to services by service providers and clients respectively.

The multi-hop routing nature of ad hoc network has seemed to be promising to unplanned circumstances like natural disasters, emergency systems, law enforcement... where the infrastructure for communication may have been destroyed or inexistent. Ad hoc network has the potential of a prominent tool for disseminating information efficiently when coordinating operations in a multi-disciplinary emergency system. This has lead to a new generation of application software and protocols to emerge and has been an active area of research since the past five years. Some researchers [1] have brought routing protocols at the level of the user space to simplify the development, deployment and portability of applications.

However most routing protocols designed for ad hoc network focused on data transmission rather than data access. It can be concluded that the usage of the limited resources, available in mobile devices, are not optimized with respect to the data and services that were accessed. As such, a number of cache schemes emerged so as to generally improving the performance and network lifetime by shortening the response-time of queries and limiting multi hop transmissions to a close cache respectively.

1.2. Motivation

Despite the existence of scalable routing protocols for ad hoc network, nodes' requests in a large ad hoc network still experience long delays. This is mainly due to its inherent characteristic of multi hops path with respect to distant sources. Most of the data received are cached locally by the applications to improve performance. However, due to the affinity of nodes' requests in a particular region it is very much likely that a node requests the same object at the same quality level or in some cases at a lower or higher quality. Examples of such objects are local news and weather information with relevant images or map details for navigational purposes. The requested object may be fully or partially cached already by its neighbouring nodes or by nodes which are much closer than the source provider may be. However, existing caching methods do not consider (1) the quality of the required object, (2) how existing cached objects can contribute to fully/partially fulfil similar requests, (3) cache admission and replacement optimizations by keeping a lower quality of the cached object rather than deleting. It is to be noted that localised fulfilment of requests favours the scalability and lifetime of a mobile ad hoc network.

2. RELATED WORK

2.1. Scalability of routing protocols

Most protocols designed for ad hoc networks were not meant for large scale networks. A few routing protocols that have the characteristics of scaling up are MLANMAR [2], CAMP [3], RBM [4], LAM [5], PBM [6], HSR [7], FSR [8],MHMR [9], SENCAST [10] and WSR [11]. The properties of these protocols which contributed to their scalability are (1) reducing the frequency of update messages, (2) reducing the size of the update messages as far as possible, (3) the maintenance of link failures are localized, (4) locality of communication as the network grows, (5) cluster-based routing approach with cluster-head, and (6) GPS assisted routing, which tremendously reduces the number of packets sent for route discovery. It has also been shown by [12] that delay-tolerant applications can take the advantage of nodes' mobility to scale up.

As the network becomes larger, the following should be noted as well: (1) it is obvious that whenever the data source is far from the requester, the latter will inevitable experience long query delays due to the multi-hops path, (2) The network may also exhibit low data availability whenever one of the node along the path fails to respond, (3) affinity of nodes requests, based on actual locations, may waste bandwidth and eventually reducing the network lifetime, and, (4) just like the notion of localized maintenance is beneficial, so will be the fulfillment of requests as far as possible.

In [13, 14], it has also been shown that sharing of information by protocols across different layers increases network performance. A number of well-known protocols were studied in [15] to show the scalability of routing protocols for heterogeneous and homogeneous MANETs.

2.2.   Cache replacement schemes for ad hoc network

Apart from a having a faster response to queries, caching can be used to increase the lifetime of the network as shown in [16] by reducing energy consumption. A number of caching replacement schemes were evaluated in [17, 18, 19] and are summarised as follows:

- LRU policy - Least recently used objects are swapped first.

- LRUMIN policy – since objects may be of different sizes, LRUMIN minimizes the number of objects to be replaced. As such LRUMIN leaves generally smaller objects in the cache for a longer period.

- Size policy – the largest object is replaced with the newly accessed object. Ties, if occurred, are resolved using time since last accessed.

- Key-based policies – objects' attributes like frequency of access, size, time since last access and entry time in cache are the keys determining the replacement. Objects to be replaced are sorted and swapped using different levels of keys.

- Function-based replacement policies - apply a sort of formula that involves a number of objects' attributes, as stated above, to select the object to be disposed.

In addition to a cache replacement scheme, a proper admission policy is also a requirement for effective caching. With respect to the characteristic that ad hoc networks are resource-scarce, it may prove ineffective to cache all objects that are accessed. In [20], it was proposed a cooperative cache-based data access framework to let mobile nodes cache the data, data path or both. Also in [21], the concept of intermediate nodes fulfilling requests by checking local cache was showed without supplementing any admission control or cache replacement scheme.

3. SAMPCAN

3.1.   SAMPCAN Communication Model

The SAMPCAN communication model is shown in figure 1. The novel caching technique proposed is SAMPCAN which is based on subsampling data for a more efficient caching in large ad hoc networks. Downsampled versions of the required object come from several caching nodes to build up the requested data.

Client applications request data via their local proxy which checks the local cache. If the request cannot be fulfilled locally, it is forwarded to neighbouring nodes which in turns forward the request if the latter is unable to contribute to the reply. Intermediate nodes, along the path taken by the request, respond to the client node when they can fully/partially contribute to the reply on behalf of the information source, i.e. the server. The partial replies from the intermediary nodes are then upsampled by the client proxy to fulfil the initial request.

In the worst case, the request reaches the server which replies the sender. Nodes along the returning path may choose to cache the reply fully/partially depending on the outcome of the local cache admission control.

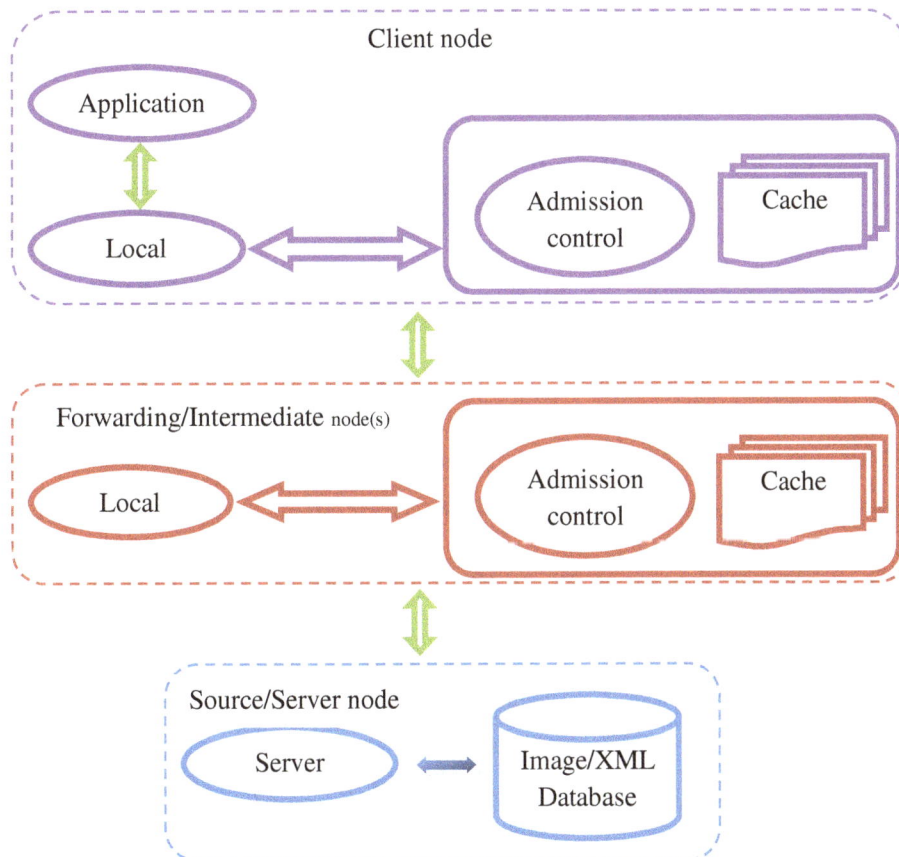

Figure 1. SAMPCAN communication model

3.2.   Data types supported

The proposed caching method allows for the efficient transmission of mainly two data types associated with location-based information (1) XML Document, and (2) Images.

3.2.1. XML Document

Database records in XML format as XML was designed to transport and store data. For example, information from an XML weather service may look like table 1.

Table 1. XML weather data example.

```
<?xml version="1.0?>
<weather_observations>
        <location>New York</location>
        <latitude>40.56</latitude>
        <longitude>-74.54</longitude>
        <weather>Sunny</weather>
        <temp_celcius>28.4</temp_celcius>
</weather_observations>
```

The above can be consumed by any application for a suitable display.

3.2.2. Images in jpeg 2000 format, for places of interests & map

The JPEG 2000 image format is preferred mainly because (1) images encoded JPEG 2000 with multiple quality layers are more error-resilient than JPEG as shown in [22], (2) downloading/streaming of jpeg 2000 images region-of-interest (ROI) is possible using HTTP, as depicted in [23]. Other interesting features are highlighted in [24].

As noted above, it is possible to build an image of higher quality when images of lower quality are available. Moreover, if different parts/tiles of an image are accessed by different nodes, a larger image or a particular ROI can be build from what was retrieved earlier.

As most applications cache whatever is retrieved, the IP layer of intermediary nodes queries the application layer before forwarding any http request. If the latter holds a valid copy of the requested information, a reply on behalf of the source is sent to the requestor.

3.3. Caching algorithm

The caching algorithm is depicted using a scenario where a node, x, requests location-based information from y:

Node x checks its local cache. Given that the required object is not present, the request is forwarded to other intermediate nodes to data source y using any georouting protocol. If the IP of y is unknown, the message is then forwarded to the location of the requested information.

Intermediary nodes check their local cache through their respective proxy. If any one of them holds a valid copy of the data requested, it sends an acknowledgement, on behalf of the source. As the reply moves from the proxy towards the requestor, depending on the routing protocol used, intermediary nodes' addresses are piggybacked with the acknowledgement message. Alternatively, georouting may be used, instead of stacking a set of IP addresses to maintain a path, to increase efficiency.

The client node chooses the closest proxy for the data. Depending on the outcome of the admission control and replacement policies, discussed in the next section, intermediary nodes may choose to save a copy of the forwarded object. In order to avoid localized cache redundancies, i.e. close neighbours caching the same pieces of data, the cache control of intermediary nodes ensures that the proxy lies in a different track and sector as depicted in the figure 2. As the number of tracks decreases: (1) cache redundancies across neighbouring nodes increases, (2) resiliency increases, and (3) frequency of updating cache records increases as the probability that the proxy lies in a different track increases.

The distance between any node and the information source/server, calculated using the standard two points on a 2D plane formula using the conversion from geodetic to 2D coordinate system using earth model in [25], determines the track. The GPS coordinates of the receiver or caching node with that of the source determines the sector as shown next.

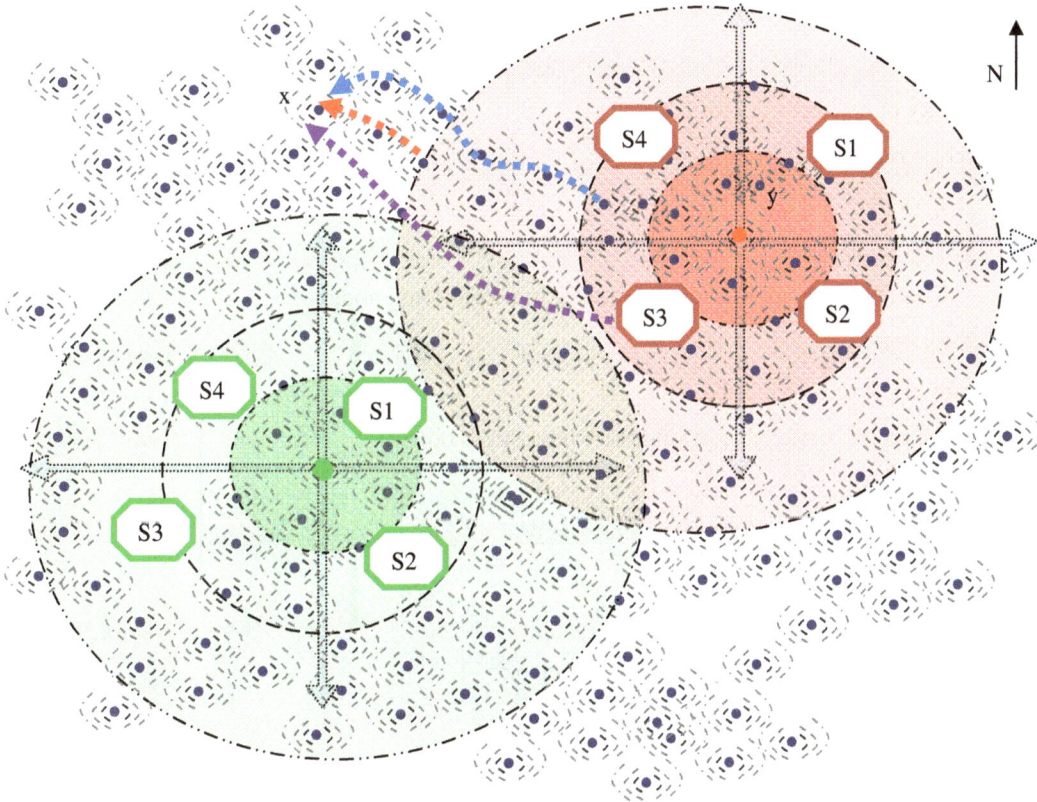

Figure 2. Track and sectors

Given that the longitude and latitude of the source are α and β respectively and four sectors are chosen, a receiver/caching node φ is in:

Sector S1, if longitude(node φ) is > α and latitude(node φ) is > β
Sector S2, if longitude(node φ) is > α and latitude(node φ) is < β
Sector S3, if longitude(node φ) is < α and latitude(node φ) is < β
Sector S4, if longitude(node φ) is < α and latitude(node φ) is > β

3.4.   Pseudo codes

Node x requesting for an object from location y

<u>Step 1 - Node x:</u>

Request $object_x$ from location y

Local proxy checks the local cache

If $object_x$ is found

> If valid

>> Use $object_x$

> Else delete $object_x$ & forward request to neighbouring nodes

Else forward request to neighbouring nodes, goto step 2

<u>Step 2 - Request forwarded by intermediary node n_i:</u>

Local proxy of n_i checks the local cache

If $object_x$ is found

> If valid

>> Piggybacked $object_x$ with reply to node x

> Else delete $object_x$ & forward request to neighboring nodes

Else

Forward request to neighboring nodes

Update requestMap, set value to value+1 where key = $object_x$

<u>Step 3 - Data source from location y sends reply:</u>

$object_x$ is sent to node x

<u>Step 4 - Reply forwarded by intermediary node n_i:</u>

Forward reply towards node x

Local proxy checks actual track and sector wrt to location y

If sender of $object_x$, or subsample of $object_x$, is located in a different track and sector

Run cache replacement function, step 5

(Nb sender can be source of information or a proxy.)

<u>Step 5 - Cache replacement function – image / XML data:</u>

Set TTL = 0

While free space < Space required for $object_x$ or its subsample

>   Increment TTL

For i up to number of objects in cache

    If time-to-live of $Object_i$ is equal to 0, remove $Object_i$

    Else If value in requestMap* related to key $object_x$ > No. of requests for $Object_i$ in cache

    OR time-to-live of $Object_i$ is <= TTL

        Replace $Object_i$ with subsample($Object_i$)

Save $object_x$ or its subsample

*The request map stores the number of requests received for a particular object (an image or a database/records).

3.5. Object sub-sampling methods for transmission/storage

3.5.1. Image sub-sampling

Objects are basically downsampled to use up less storage space when a lower quality of the former is still fulfilling its purpose. Given that four nodes at different locations retrieved a subsampled $object_x$, these subsampled versions, when cached, are useless when a request for a better quality of the object emerges from any nodes. However, the cached objects of lower quality can be used to reconstruct objects of better quality when the sender uses different offset values and sample periods to generate subsampled objects. The following illustrates:

Given that an image, img, is as shown in figure 3, whose the starting corner of the image is identified by x and y and (2) the number of samples that exist is υ, where there υ is equal to the sample period along the x axis multiply by the sample period along the y axis, then the subsample a is defined by:

x = offset of img for sample a along the x-axis

y = offset of img for sample a along the y-axis

Set i and j to zero

for i is less than img.length

      for j is than img.width

          sample a = sample a + img.pixel(x+i, y+j)

          increment j by the sample period

      increment i by by the sample period

      Reset j to zero

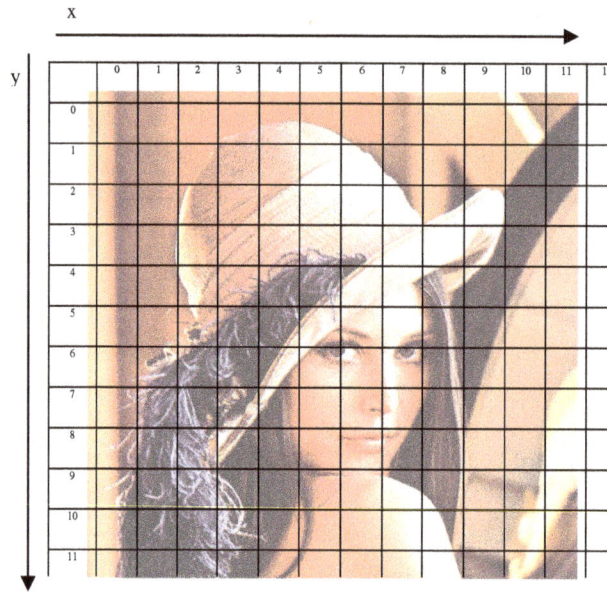

Figure 3. Subsampling an image given an offset and a period

Figure 4 and figure 5 show the image cells/pixels chosen for two samples (a & b) respectively when the above algorithm is applied to img with sample period 2 for both the x-axis and the y-axis (here $\upsilon=4$), and the subsample offset (x, y) for sample a, b, c and d are (0, 0), (1, 0), (0, 1) and (1, 1):

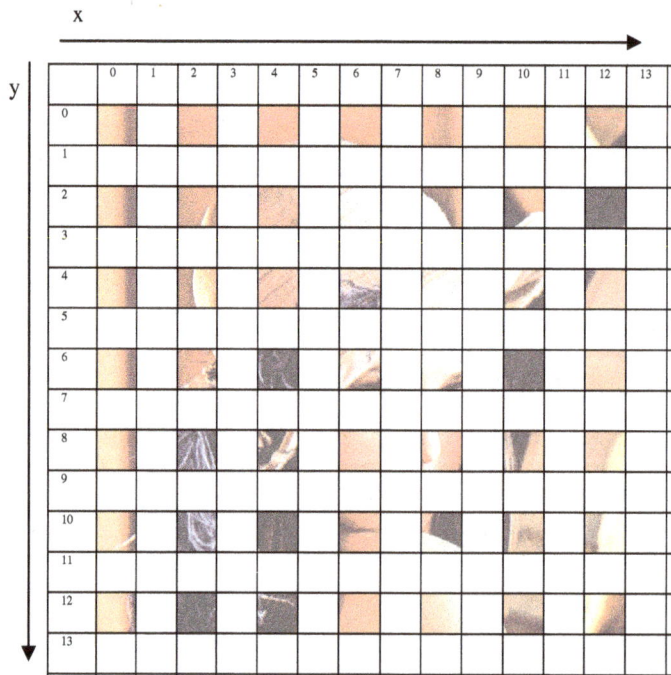

Figure 4. Image cells/pixels used for sample a

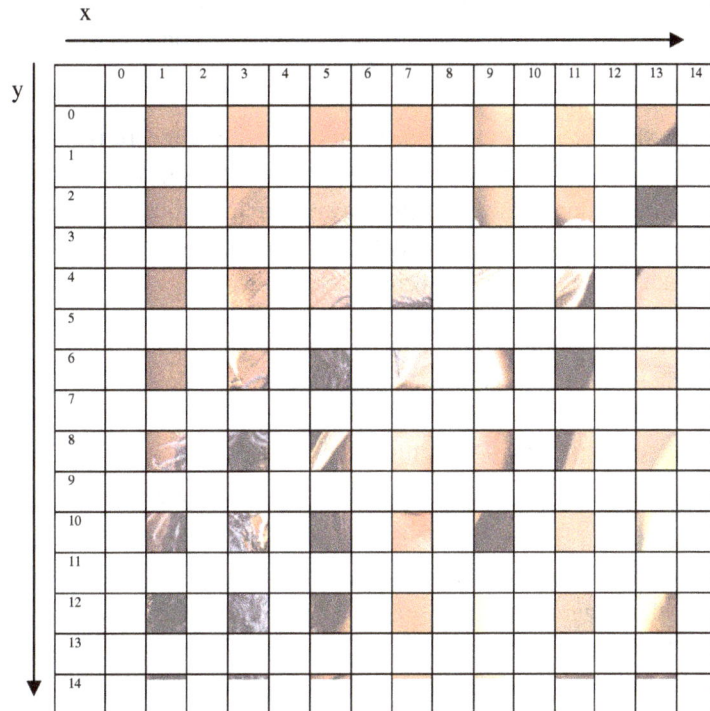

Figure 5. Image cells/pixels used for sample b

Note that the Subsampling period and offset indicate the number of pixel to advance in a direction after choosing a pixel for the sample and the first pixel of the subsample image respectively.

3.5.2. XML data sub-sampling

Records accessed by nodes are stored in XML forms. The root element and the first level branch have the name of the database and table respectively. For example, table 2 shows data that was accessed from the table **observation** found in the database **weather_observations** by two different nodes, x and y.

Table 2. Example of xml data for weather observations.

`<?xml version="1.0?>`
`<weather_observations>`
`<observation>`
`<rec_id>1</rec_id>`
`<location>New York</location>`
`<latitude>40.56</latitude>`
`<longitude>-74.54</longitude>`
`<weather>Sunny</weather>`
`<temp_celcius>28.4</temp_celcius>`
`<obserTime>16 :00</obserTime>`
`</observation>`
…
`</weather_observations>`
	`<?xml version="1.0?>`
	`<weather_observations>`
	`<observation>`
	`<rec_id>2</rec_id>`
	`<location>New State</location>`
	`<latitude>45.46</latitude>`
	`<longitude>-34.54</longitude>`
	`<weather>Cloudy</weather>`
	`<temp_celcius>18.4</temp_celcius>`
	`<obserTime>16 :10</obserTime>`
	`</observation>`
	…
	`</weather_observations>`

Node x	Node y

[26] brought forward XQuery to manipulate xml data like database data. As such details of weather observations are easily retrieved from intermediate nodes and merged to fulfill other queries.

Subsampling of an xml document can be done at the level of a record tag, e.g. 'observation', or an attribute/field of a record, e.g. 'location'. In the latter case, the attributes of a record are listed in order of priority/importance, then the downsampling process deletes the last attribute of the record as listed. When node x downsamples the 'weather_observations' document, in a first instance the attribute 'obserTime' of all records gets deleted.

Whenever location information is available, like in the above file, higher priority is given to caching records whose locations is nearest to the node's actual position

4. DISCUSSIONS/EXPERIMENTATION

Java Advanced Imaging [27] was used to manipulate the JPEG 2000 image. Images were downsampled and used to reconstruct their originals as shown in the series of figures below. Using $\upsilon=4$, downsampled versions (a, b, c & d) of the original lena image were obtained as shown in figure 6. The original image is of size 349 Kb and the downsampled image size given $\upsilon=4$ was 109 Kb. However, there is no direct relationship between the downsampled image size and υ, as images' attributes largely affect its compression. Figure 7 and 8 shows reconstructed images when subsamples a & b and subsamples a,b & c are available respectively. While figure 9 shows the original image reconstructed given the availability of all four subsampled versions.

Whenever images were reconstructed, missing pixel information was considered as white. An improvement is sought with a block-based predictive algorithm to predict the missing pixel information using the values of immediate neighbours.

Despite the overall downsampled data sent is relatively larger than the original, there is still a reduction the risk of network partitioning and an increase the lifetime of the network as the data comes from (1) various parts of the network, and (2) proxies that are closer than the actual source would be.

It should also be noted that this novel caching technique can be applied to any other client-server model with multi-hop paths.

Lenasubsample a – Size: 109 Kb, resolution: 256 * 256	Lenasubsample c – Size: 109 Kb, resolution: 256 * 256
Lenasubsample b – Size: 109 Kb, resolution: 256 * 256	Lenasubsample d – Size: 109 Kb, resolution: 256 * 256

Figure 6. Subsampling of Lena image with $\upsilon=4$

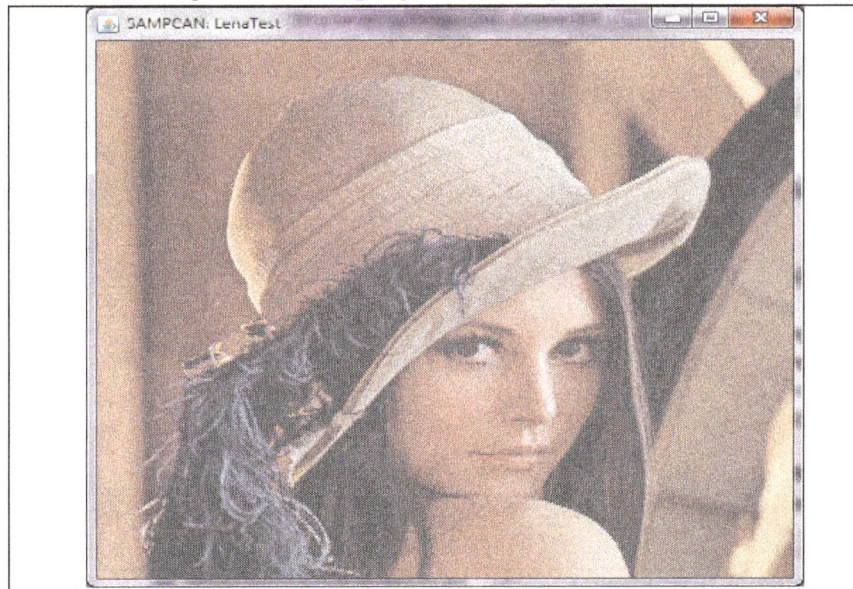

Figure 7. Reconstructed image – given sample a & b, resolution: 512 * 512

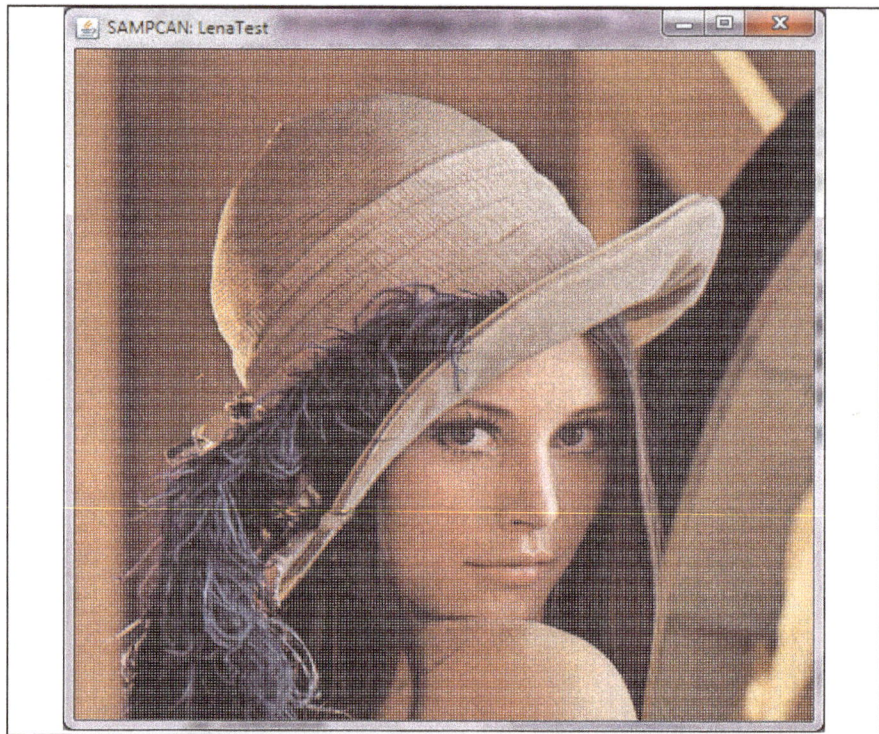

Figure 8. Reconstructed image – given sample a, b & c, resolution: 512 * 512

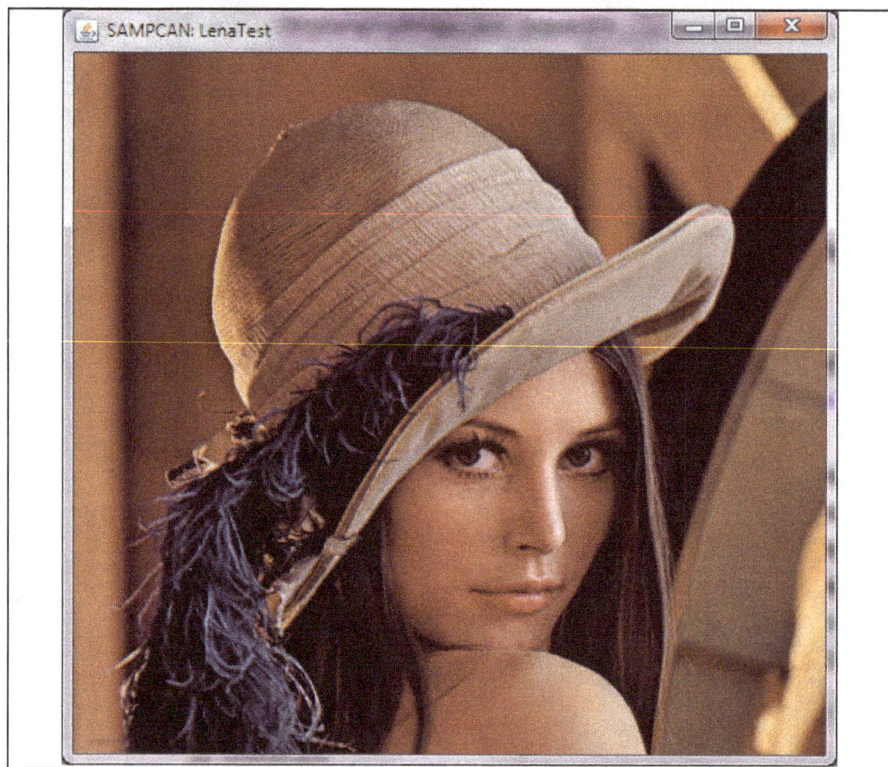

Figure 9. Reconstructed image – given all four downsampled versions, resolution: 512 * 512

5. FUTURE WORKS

A context-aware cache admission control is required for to dynamically update the threshold of the following parameters which can be used to optimise the replacement of a cached object c_x at node n_i: (1) distance between n_i and the data source x, (2) object time-to-live as timestamped by the data source, (3) Number of similar requests, (4) time of last request, and (5) whether subsample of the object exists already. The optimum number of tracks and sectors with respect to context of use needs to be evaluated before applying SAMPCAN.

6. CONCLUSION

One of the mentioned features for the scalability of routing protocols is the locality of communication as the ad hoc network grows. This mode of interaction was restricted to nodes in the process of maintaining communication paths or gathering local context information. In the same line, SAMPCAN promotes localised communication in the interactions of client and server dimension. SAMPCAN introduces a novel caching technique that optimised the caching storage using subsampling that can support mainly two data types, images and XML document. Cached elements are accessible via a local proxy by neighbouring nodes requesting similar data. The quality of objects required can be augmented using subsampled cached versions from several nodes. Apart from extending the availability of the local cache to neighbouring nodes, the caching method optimises storage with at least maintained cache hit ratio as existing objects are not swapped with new elements but downsampled.

7. REFERENCES

[1]   Pedro García López, Raúl Gracia Tinedo and Josep M. Banús Alsina, Moving routing protocols to the user space in MANET middleware, Journal of Network and Computer Applications, vol. 33, issue 5, 2010.

[2]   Yi Y., Gerla M. and Obraczka K., "Scalable team multicast in wireless ad hoc networks exploiting coordinated motion", Ad Hoc Networks Journal, August 2003.

[3]   Madruga E. L. and Garcia-Luna-Aceves J. J, "Scalable multicasting: The core-assisted mesh protocol", ACM/Baltzer Mobile Networks and Applications, Special Issue on Management of Mobility, 6(2):151--165, 2001.

[4]   Corson M. S. And Batsell S. G., "A reservation-based multicast (RBM) routing protocol for mobile networks: Initial route construction phase", ACM/Baltzer Wireless Networks, vol. 1, no. 4, pp. 427-450, Dec 1995.

[5]   Ji L. and Corson M. S., "A lightweight adaptive multicast algorithm", Proc. of Globecom'98, 1998.

[6]   Mauve M., Füßler H., Widmer J. and Lang T., "MobiHoc Poster: Position-based multicast routing for mobile ad hoc networks", Mobile Computing and Communication Review, Vol. 7, No. 3, pp. 53 – 55, 2003.

[7]   Iwata A., Chiang C., Pei G., Gerla M. and Chen T., "Scalable routing strategies for ad hoc wireless networks", IEEE Journal on Selected Areas in Communications SAC, 1999, Vol. 17, pp. 1369-1379.

[8]   Pei G., Gerla M. and Chen T., "Fisheye State Routing: A Routing Scheme for Ad Hoc Wireless Networks", IEEE International Conference on Communications, 2000, Vol 1, pp. 70-74.

[9]   An B. and Papavassiliou S., "A mobility-based hybrid multicast routing in mobile ad hoc wireless networks", Proc. of MILCOM 2001, Vienna, VA, Oct 2001.

[10]   Appavoo P. and Khedo K., "SENCAST: A Scalable Protocol for Unicasting and Multicasting in a Large Ad hoc Emergency Network", International Journal of Computer Science and Network Security, Vol.8 No.2, February 2008.

[11]   Acer U.G., Kalyanaraman S. and Abouzeid A.A., Weak State Routing for Large-Scale Dynamic Networks, IEEE/ACM Transactions on Networking, vol. 18, issue 5, 2010.

[12]   Grossglauser M. and Tse D. N. C., "Mobility Increases the Capacity of Ad hoc Wireless Networks", IEEE/ACM Transaction on Networking, vol. 10. no. 4,Aug. 2002.

[13]   Chen K., Shah S. H. and Nahrstedt K., "Cross-Layer Design for Data Accessibilityin Mobile Ad Hoc Networks", Wireless Personal Communications 21: 49–76, 2002.

[14]   Conti M., Maselli G., Turi G. and Giordano S., "Cross-Layering in Mobile Ad Hoc Network Design", Computer IEEE Computer Society, pp 48, 2004.

[15]   Al Amri H., Abolhasan M. And Wysock T., Scalability of MANET routing protocols for heterogeneous and homogenous networks, Signal Processing and Communication System, Elsevier, vol. 36, issue 4, 2010.

[16]   Nitnaware D. and Verma A, Energy constraint Node cache based routing protocol for Adhoc Network, International Journal of Wireless & Mobile Networks (IJWMN), Vol.2, No.1, 2010.

[17]   Abrams M., Standridge C., Abdulla G., Williams S. and Fox E., "Caching Proxies: Limitations and Potentials," Proc. Fourth Int'l World Wide Web Conf., Boston, 1995.

[18]   Aggarwal C., Wolf L. and Yu P., "Caching on the World Wide Web", IEEE Transactions on Knowledge and Data Engineering, vol. 11, no. 1, 1999

[19]   Williams S., Abrams M., Standridge C. R., Abdulla G., and Fox E. A., "Removal Policies in Network Caches for World Wide Web Documents," Proc. ACM SIGCOMM, pp. 293-304, 1996.

[20]   Cao G., Yin L. and Das C. R., "Cooperative Cache-Based Data Access in Ad Hoc Networks", Computer – IEEE Computer Society, 2004, VOL 37; PART 2, pages 32-39.

[21]   Gonzalez-Canete F. J., Casilari E, and Trivino-Cabrera A., Proposal and evaluation of a caching scheme for ad hoc networks, ADHOC-NOW '09 Proceedings of the 8th International Conference on Ad-Hoc, Mobile and Wireless. 2009

[22]   Pekhteryev G., Sahinoglu Z., Orlik P. and Bhatti G., "Image Transmission over IEEE 802.15.4 and ZigBee Networks", Mitsubishi Electric Research Laboratories, 2005.

[23]   Deshpande S., Zeng W., "Scalable Streaming of JPEG2000 Images using Hypertext Transfer Protocol", 2001.

[24]   Athanassios S., Charilaos C., and Touradj E., The JPEG 2000 Still Image Compression Standard, IEEE Signal Processing Magazine, 2001.

[25]   Carlson C. G. and D. E. Clay D. E., "The Earth Model – Calculating Field Size and Distances between Points using GPS Coordinates", Site-Specific Management Guidelines series-11, Potash & Phosphate Institute (PPI), 1999.

[26]   W3C XML Query (XQuery), http://www.w3.org/XML/Query/, accessed 1st February, 2011.

[27]   Oracle Corporation, Java Advanced Imaging (JAI) API, http://java.sun.com/javase/technologies/desktop/media/jai/, accessed 13th November, 2010.

5

Efficient Spectrum Sharing in the Presence of Multiple Narrowband Interference

Demosthenes Vouyioukas

Department of Information and Communication Systems Engineering
University of the Aegean
Karlovassi 83200, Samos, Greece

e-mail: dvouyiou@aegean.gr

Abstract. *In this paper, we study the spectrum usage efficiency by applying wideband methods and systems to the existing analog systems and applications. The essential motivation of this work is to define the prospective coexistence between analog FM and digital Spread Spectrum systems in an efficient way sharing the same frequency band. The potential overlaid Spread Spectrum (SS) system can spectrally coincide within the existing narrowband Frequency Modulated (FM) broadcasting system upon several limitations, originating a key motivation for the use of the FM radio frequency band in many applications, encompassing wireless personal and sensors networks. The performance of the SS system due to the overlaying analog FM system, consisting of multiple narrowband FM stations, is investigated in order to derive the relevant bit error probability and maximum achievable data rates. The SS system uses direct sequence (DS) spreading, through maximal length pseudorandom sequences with long spreading codes. The SS signal is evaluated throughout theoretical and simulation-based performance analysis, for various types of spreading scenarios, for different carrier frequency offset (Δf) and signal-to-interference ratios, in order to derive valuable results for future developing and planning of an overlay scenario.*

Key-Words: *Spread Spectrum, Direct Sequence (DS), Analog FM, Interference, Bit-Error-Rate.*

1. Introduction

Nowadays and future trends of wireless systems are mainly towards developing new wireless technologies so as to support higher data rates with minimum costs. This is achievable – apart from other – by demanding additional access to the spectrum. The spectrum scarcity is the most serious challenge facing the wireless industry today and it is only going to get worse during the next years. Major accomplishments have already been developed in the area of Cognitive Radio (CR) for spectrum sensing, spectrum sharing, dynamic spectrum access to primary and secondary users and between Radio Access Technology (RAT) [1],[2]. But, spectrum regulation still remains where many spectrum bands are unused and some of them are heavily congested allocated to specific applications and operators.

Enabling the provision of services and applications in an efficient way making use of the natural resource radio spectrum can be realized by introducing principles of wideband SS techniques in real existing regulation. The spectrum overlay between Direct Sequence Spread Spectrum (DSSS) systems and FM radio broadcasting services can increase spectral efficiency and communication capacity within a geographical area. Anti-jamming, anti-interference, privacy and low power spectral density are some of the advantages that the SS technique encompasses and strengthen their use to coexisting schemes. Both field tests and analyses have provided a perspective as to what the capabilities of such system are.

The mutual interference, which in specific conditions of high density becomes unavoidable, the estimation of interference noise serves as a measure for accepting their coexistence. This overlay concept has been demonstrated in both PCS band [3] and cellular band [4]. In the proposed work of [5], it has been proposed a CDMA network that is permitted to be overlaid on top of existing microwave narrow-band users that occupy a part of the spread-spectrum system bandwidth (BW). In fact, this is another advantage of CDMA over either TDMA or FDMA, because this will increase the overall spectrum capability. However, such an application must be considered carefully, because the CDMA (wide-band) users and the narrow-band users can interfere with each other. Analysis of Ultra-Wide-Band (UWB) systems reveals that signals are suitable for underlay communications and the design of these systems are very promising, where it is necessary the understanding of the effects of interference to and from narrow-band systems [6].

Simulcasting methods and techniques [7],[8] have been developed for simultaneous transmission of digital data with analog FM, over frequency bands in the 87.5 – 108 MHz range. Some of these new methods are compatible with the standards of analog FM for terrestrial broadcast, while others are designed for new frequency bands. The method of hybrid in band on channel (HIBOC) has been used, in order to transmit digital audio simultaneously with analog FM, using frequency-orthogonal analog and digital transmission over a total system bandwidth of 400 kHz.

Recent advances in wireless communications and electronics have enabled the development of various types of short-range wireless communication systems, such as WLANs, Bluetooth, High data rate WPANs, Low power WPANs, Body area networks (BANs) and RFID-technologies with a variety of applications (e.g., commercial, home, health, military, etc.). Most of these systems as well as the wireless sensors networks, utilize the spread spectrum technique for the physical layer [9]. The unusual application requirements of sensor networks make the choice of transmission media more challenging. One option for radio links is the use of industrial, scientific and medical (ISM) bands, which offer license free communication in most countries. These frequency bands suffer from limitations and harmful interference from existing applications and applications that may emerge in the future [10]. Another option for radio links is the use of the already occupied frequencies, in an overlay scenario utilizing the ultra low power transmission of the sensor nodes.

Exploitation of this spectrum overlay concept in the physical layer in the field of wireless personal and sensor networks, can increase communications capacity and spectral efficiency, but may cause the following types of interference: i) interference from the narrowband FM stations to the SS system and ii) interference from the overlaid wideband SS system on the FM receivers. The first is the scope of this paper. The later case was investigated and the correspondent results were presented in [11].

The performance of the Direct Sequence Spread Spectrum system is examined once overlaid on top of the existing radio band of multiple FM stations. The probability of error for various FM interfering conditions is derived for the case of single and multiple spread spectrum users. The simulation of the proposed configuration for the estimation of the error rate, aims for the best approach of a realistic system, as well as for the knowledge of the produced signals at the output of each module. Consequently, based on the analysis and according to mathematic analytical models of each module, the simulation of the whole system is realized, taking into consideration all the stimulated occurrences that are caused due to non-ideally operation of the passive and active elements. Moreover, the Additive White Gaussian Noise (AWGN) is taken into account for the Spread Spectrum system and also for each FM station.

In Section 2, the Spread Spectrum system is briefly described in the form used in this paper. The total interference analysis for single and multiple SS due to multiple narrowband interference users is performed in Section 3, while the FM power spectrum approximation is presented in Section 4 along with some of their properties in the context of closed-form mathematical expressions. Section 5 contains some indicative numerical results for single and multiple SS users, while in Section 6 simulation results for a single SS user are presented. Finally, a brief discussion and concluding remarks are given in Section 7.

2. Description of Spread Spectrum System

The spread spectrum signal is given by [12] as

$$s(t) = \sum_{k=1}^{K} s_k(t - \tau_k) = \sum_{k=1}^{K} \sqrt{2P_k} b_k(t - \tau_k) c_k(t - \tau_k) \cos(\omega_o t + \theta_k) \tag{1}$$

where P_k is the received power of the k-th spread spectrum, τ_k is the time delay of the signal uniformly distributed into [0,T], θ_k is the phase angle uniformly distributed on [0,2π], $b_k(t)$ is the modulating digital signal of the k-th user given by $b_k(t) = \sum_{n=-\infty}^{\infty} a_n^{(k)} p_{T_b}(t - nT_b)$ where $\{a_n = \pm 1, -\infty < n < +\infty\}$ and p_{T_b} is the rectangular pulse of T_b duration, with $P[b_k(t)=1]=P[b_k(t)=-1]=0.5$, and ω_o is the carrier frequency of the signal. $c(t)$ is the spreading code given by $c_k(t) = \sum_{n=-\infty}^{\infty} c_n^{(k)} p_{T_c}(t - nT_c)$ where $c_k(t) \in \{\pm 1\}$ is one chip of random binary sequence $\{c_n\}$, which consists of independent identically distributed (i.i.d) random variables with equal probability, T_c is the spreading code chip duration and p_{T_c} is the chip waveform, assumed rectangular.

3. Interference Analysis

3.1   Single SS user

We consider the receiver model as depicted in Figure 1. The total received signal is corrupted by noise and interference

$$r(t) = s(t) + i(t) + n(t) \tag{2}$$

where $i(t)$ denotes the interference and $n(t)$ is the zero-mean white Gaussian noise. Assuming that code synchronization has been established and we have perfect channel estimation, the input to the demodulator is

$$r_1(t) = s(t)c(t) + i(t)c(t) + n(t)c(t) \tag{3}$$

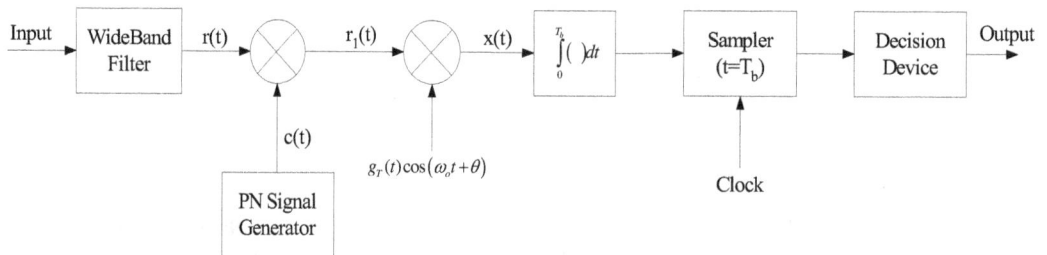

Figure 1: DS receiver model with BPSK modulation

The input quantity to the decision device, during the period T_b of the data signal, is

$$Z_o = \int_0^{T_b} r_1(t)\cos(\omega_o t+\theta)dt = \int_0^{T_b}\left[\sqrt{2P}b(t)c(t)\cos(\omega_o t+\theta)+i(t)+n(t)\right]c(t)\cos(\omega_o t+\theta)dt =$$

$$= \sqrt{2P}\int_0^{T_b} b(t)c^2(t)\cos^2(\omega_o t+\theta)+\int_0^{T_b} i(t)c(t)\cos(\omega_o t+\theta)dt+\int_0^{T_b} n(t)c(t)\cos(\omega_o t+\theta)dt$$

(4)

We assumed that $f_o = \omega_o/2\pi \gg 1/T_b$, so that the double frequency term $\left(\cos^2(\omega_o t+\theta)\cong 1+\cos(2\omega_o t)\cong 1\right)$ is negligible, suppressed by the bandpass filter following the demodulator. From equations (2), (3), (4) and the fact that $b(t)$ is constant over the period T_b we have

$$Z_o \cong \sqrt{2P}\int_0^{T_b} b(t)\sum_{n=0}^{G_p-1} p_{T_c}^2(t-nT_c)dt + S_i + S_n = \pm\sqrt{2P}T_b + S_i + S_n, \quad b(t)\in\{\pm 1\} \quad (5)$$

where

$$\int_0^{T_b}\sum_{n=0}^{G_p-1} p_{T_c}^2(t-nT_c)dt = G_p T_c = T_b \tag{6}$$

$$S_i = \int_0^{T_b} i(t)c(t)\cos(\omega_o t+\theta)dt \tag{7}$$

$$S_n = \int_0^{T_b} n(t)c(t)\cos(\omega_o t+\theta)dt \tag{8}$$

The code chip sequence $\{c_n\}$ can be modeled as a random binary sequence, comprises by statistically independent symbols with equal probability, is uncorrelated (white) and therefore $E[c_n c_m] = E[c_n]E[c_m]$ for n≠m and these conditions imply that $E[c_n]=0$ and $E[c_n^2]=1$. Perfect code, phase and symbol synchronization are assumed.
Substituting the spreading code into (7), we obtain

$$S_i = \sum_{n=0}^{G_p-1} c_n J_n \tag{9}$$

where

$$J_n = \int_{nT_c}^{(n+1)T_c} i(t)p_{T_c}(t-nT_c)\cos(\omega_o t+\theta)dt \tag{10}$$

Likewise

$$S_n = \sum_{n=0}^{G_p-1} c_n N_n \tag{11}$$

where

$$N_n = \int_{nT_c}^{(n+1)T_c} n(t)p_{T_c}(t-nT_c)\cos(\omega_o t+\theta)dt \tag{12}$$

The equation (5) can also be written as $S \cong S_o + y_i$ where S_o is the desired signal and $y_i = S_i + S_n$ denotes the total additive interference plus Gaussian noise. The first term of the right-hand side of the equation is deterministic and its value is given by equation (5). The terms of S_i are independent random variables with zero mean value uniformly bounded and

$\text{var}\{S_i\} \to \infty$ with $G_p \to \infty$, so applying the Central Limit Theorem (CLT) [14] implies that S_i converges in a Gaussian distribution with zero mean value and variance 1. Consequently, the distribution of S_i is almost Gaussian when G_p is large. We must underline, that for the present analysis long sequences have been utilized with large processing gain. Therefore, the second term is a summation of two zero-mean independent Gaussian variables, indicating that y_i will have an almost zero-mean Gaussian distribution and variance given by $\text{var}\{y_i\} = \text{var}\{S_i\} + \text{var}\{S_n\}$. Based on CLT and Gaussian approximation, the probability of error depends on the statistical characteristics of the interference. The final probability of error will be given by [14]

$$P_e = \frac{1}{2}\text{erfc}\left[\sqrt{\frac{E_b T_b}{\text{var}\{S_i\} + \text{var}\{S_n\}}}\right] \tag{13}$$

and the total signal-to-noise ratio at the output of the correlator by

$$SNR_D = \frac{E^2[S]}{\text{var}\{S_i\} + \text{var}\{S_n\}} \tag{14}$$

where $E_b = PT_b$ is the received bit energy $b(t)$, P is its mean power and *erfc* is the complementary error function. Based on the previous definitions, the variance of the variable y_i is

$$\sigma_{y_i}^2 = \text{var}\{y_i\} = \text{var}\{S_i\} + \text{var}\{S_n\} = E\left[S_i^2\right] + E\left[S_n^2\right]$$
$$= G_p E\left[J_n^2\right] + G_p E\left[N_n^2\right] \tag{15}$$

Suppose that θ is an independent random variable uniformly distributed in over $[0, 2\pi]$. The stationary of $n(t)$ and a change of variables implies that the variance of N_n is

$$E\left[N_n^2\right] = \frac{N_o T_c}{4} \tag{16}$$

where $\dfrac{N_o}{2}$ is the two-sided noise power spectral density (PSD) and we assumed that $f_o = \omega_o / 2\pi \gg 1/T_c$.

For the calculation of the variance of interference, following the noise procedure results to [13]

$$E\left[J_n^2\right] = \frac{1}{2}T_c \int_{-T_c}^{T_c} R_i(\tau) R_p(\tau) \cos \omega_o \tau \, d\tau \tag{17}$$

where $R_i(\tau)$ is the autocorrelation of $i(t)$, $R_p(\tau)$ is the autocorrelation function of the PN sequence. The limits of the integral can be extended to infinity because the integrand is truncated. Because $R_i(\tau)$ is an even function, the convolution theorem and the known *Fourier* transform of $R_p(\tau)$, yield the following [15]

$$E\left[J_n^2\right] = \frac{1}{2}T_c^2 \int_{-\infty}^{\infty} S_i(f) \sin c^2 \left[(f - f_o) T_c\right] df \tag{18}$$

where $S_i(f)$ is the PSD of the interference after passage through the wideband bandpass filter. If $S_i'(f)$ is the PSD of the interference at the input of the wideband bandpass filter and $H(f)$ is its transfer function, then $S_i(f) = S_i'(f)|H(f)|^2$. Suppose that the effects of the wideband filter and the integration over negative frequencies are negligible ($f_o \gg 1/T_c$), then

$$E\left[J_n^2\right] = \frac{1}{2}T_c^2 \int_{f_i - W_i/2}^{f_i + W_i/2} S_i(f)\sin c^2\left[(f - f_o)T_c\right]df \tag{19}$$

where we have assumed that the bandwidth of the interference is $W_i \leq W \cong 2/T_c$. The PSD of the interference $S_i(f)$ is given by equation (35) at the next section and the total probability of error is given by

$$P_e = \frac{1}{2}\text{erfc}\left[\frac{E_b}{N_o + 2\dfrac{T_b}{G_p}\displaystyle\int_{f_i - W_i/2}^{f_i + W_i/2} S_i(f)\sin c^2\left[(f - f_o)T_c\right]df}\right]^{\frac{1}{2}} \tag{20}$$

3.2 Multiple SS users

The scope of this section is to examine the performance of the Direct Sequence Spread Spectrum system in the case of multiple spread spectrum active users and to access the bit error probability in the presence of analog FM interference and multiple access interference (MAI), using the simplified improved Gaussian approximation.

We assume communication through point-to-multipoint transmission, so multiple access interference is considered with perfect code, phase and symbol synchronization, particularly for downlink transmissions.

The total received signal in the input of the desired spread spectrum receiver is

$$r(t) = \sum_{k=0}^{K-1} s_k(t - \tau_k) + i(t) + n(t) \sum_{k=0}^{K-1} \sqrt{2P_k}b_k(t - \tau_k)c_k(t - \tau_k)\cos(\omega_o t + \phi_k) + i(t) + n(t) \tag{21}$$

where $s_k(t)$ is the k-th user's transmitted signal, b_k is the data sequence for user k, c_k is the spreading sequence for user k, τ_k and φ_k are random delay and carrier phase terms relative to the desired reference user 0, P_k is the received power of user k, ω_o is the carrier frequency of the signal, $i(t)$ the interfering analogue FM signal and $n(t)$ denotes the zero mean white Gaussian noise of the channel.

The received signal contains the desired user, the K-1 undesired users and the interfering FM signal plus noise and all signals are mixed down to baseband, multiplied by the PN sequence of the desired user 0 and integrated over a bit period T_b.

The decision statistic of the receiver for user 0 is:

$$Z_o = \int_0^{T_b} r(t)c_o(t)\cos(\omega_o t + \theta)\,dt \tag{22}$$

which may be expressed as

$$Z_0 = I_0 + I_{MAI} + I_{FM} + I_n \tag{23}$$

where

$$I_o = \sqrt{2P_o}\int_0^{T_b} b(t)c_o^2(t)\cos^2(\omega_o t)\,dt \tag{24}$$

$$I_{MAI} = \int_0^{T_b}\sum_{k=1}^{K-1}\sqrt{2P_k}\,b_k(t - \tau_k)c_k(t - \tau_k)\cos(\omega_o t + \phi_k)c_o(t)\cos(\omega_o t)\,dt \tag{25}$$

$$I_{FM} = \int_0^{T_b} i(t)c(t)\cos(\omega_o t + \theta)\,dt \tag{26}$$

$$I_n = \int_0^{T_b} n(t)c(t)\cos(\omega_o t + \theta)dt \qquad (27)$$

where I_0 is the desired contribution to the decision statistic from the desired user $k=0$, I_{MAI} is the multiple access interference, MAI, from other users, I_{FM} is the contribution of the interfering overlaid FM signal and I_n is the thermal noise contribution. The contribution of the FM interference and noise term is investigated and analyzed previously (S_i and S_n respectively).

Since the performance of the Spread Spectrum system under investigation is assumed to be interference limited, which means that the FM interfering signal is the dominant interference component, the number of simultaneous users is limited (relative small K). Hence it is essential for the calculation of MAI to use an approximation, which is accurate for relative small values of K.

For the calculation of the Multiple Access Interference contribution and finally the overall performance of the system, the simplified expression of the Improved Gaussian Approximation (SIGA) is used.

3.2.1 Performance of the overlaid system based on SIGA

The simplified improved Gaussian approximation to the bit error rate is based upon the premise that the MAI converges to a Gaussian random variable as the number of chips per data bit N becomes large and for any K [16]. Based on the expansion of differences method [17], the SIGA requires only that the mean μ and the variance σ^2 of $\psi = var(I_{MAI})$ to be determined.

Based on the Central Limit Theorem and the fact that all interference components of the decision statistic are assumed to be independent random variables with distribution, which converges to Gaussian for large N, the variance of the decision statistic, is

$$\text{var}(Z_0) = \text{var}(I_{MAI}) + \text{var}(I_{FM}) + \text{var}(I_n) \qquad (28)$$

The contribution of noise and analog FM interference to the decision statistic is analyzed previously. For the case of perfect power control such that all users have identical power levels and these power levels are not random, the mean μ_ψ and the variance σ_ψ^2 of $\psi = \text{var}(I_{MAI})$, are given by [18]

$$\mu_\psi = \frac{G_p T_c^2}{6} P(K-1) \qquad (29)$$

$$\sigma_\psi^2 = (K-1)\frac{T_c^4}{4}P^2\left[\frac{23G_p^2}{360} + G_p\left(\frac{1}{20} + \frac{K-2}{36}\right) - \frac{1}{20} - \frac{K-2}{36}\right] \qquad (30)$$

and the probability of error is given by [18]

$$P_e = E\left[Q\left(\sqrt{\frac{E_b T_b}{2\left[\psi + \frac{N_0 T_b}{4} + \frac{1}{2}T_b T_c S_i(f)\right]}}\right)\right] \qquad (31)$$

The simplified bit error probability expression for the case of MAI [16], analog FM and noise interference is given by

$$P_e^{(SIGA)} \approx \frac{1}{3} erfc\left(\sqrt{\frac{E_b T_b}{4(\mu_\psi + I_{eq})}}\right) + \frac{1}{12} erfc\left(\sqrt{\frac{E_b T_b}{4(\mu_\psi + \sqrt{3}\sigma_\psi + I_{eq})}}\right)$$
$$+ \frac{1}{12} erfc\left(\sqrt{\frac{E_b T_b}{4(\mu_\psi - \sqrt{3}\sigma_\psi + I_{eq})}}\right) \tag{32}$$

where

$$I_{eq} = \text{var}\{I_n\} + \text{var}\{I_{FM}\} = \frac{N_0 T_b}{4} + \frac{1}{2} T_b T_c S_i(f) \tag{33}$$

and $S_i(f)$ is the closed-form FM power spectrum approximation given by equation (35) at the next section, where the FM power spectrum aggregation is calculated over the bandwidth of all the under-utilized FM stations.

4. FM Power Spectrum Approximation

In order to obtain accurate results for the BER of equations (20), (32) and (33), the interference power of the FM spectrum at the input of the digital receiver is required in closed-form mathematical expression. The required power spectral densities of the interference FM signals are presented as a result of laboratory measurements and mathematical derivations.

The FM signal is generally a random process and consequently can be only statistically presented. Specifically, to characterize the power spectral density of the FM signal we must measure the values of statistical parameters, such us mean, maximum, minimum and median. The measurement configuration consists of a spectrum analyzer, a biconical antenna and a laptop computer. The program in the computer controls the spectrum analyzer through HP-IB port. The spectrum analyzer parameters such as resolution bandwidth, video bandwidth, frequency span and sweep time are assigned through the Laptop PC. Three kinds of stations were recorded depending on the transmitted information; one that transmits only music, one that transmits only speech (voice) and one that transmits the combination of these two.

The statistical evaluation of the previous measurements and the derivation of a closed-form mathematical expression have been studied in [13]. Further statistical processing showed that the Gaussian fit for the estimation of the PSD of the FM channel responds very well according to several conditions of the transmitted data. The equation is given by the following expression:

$$\hat{G}_m(f) = y_o + A \exp\left[-\frac{(f - f_m)^2}{2\omega^2}\right] \tag{34}$$

where the parameters y_o is a constant near the noise threshold, A is the amplitude constant, f_m is the center frequency, ω is the standard deviation and R^2 denotes the regression analysis of the fit data. All the Gaussian fit parameters are given in Table 1, while Figure 2 depicts the mean recorded values and the Gaussian curve fit.

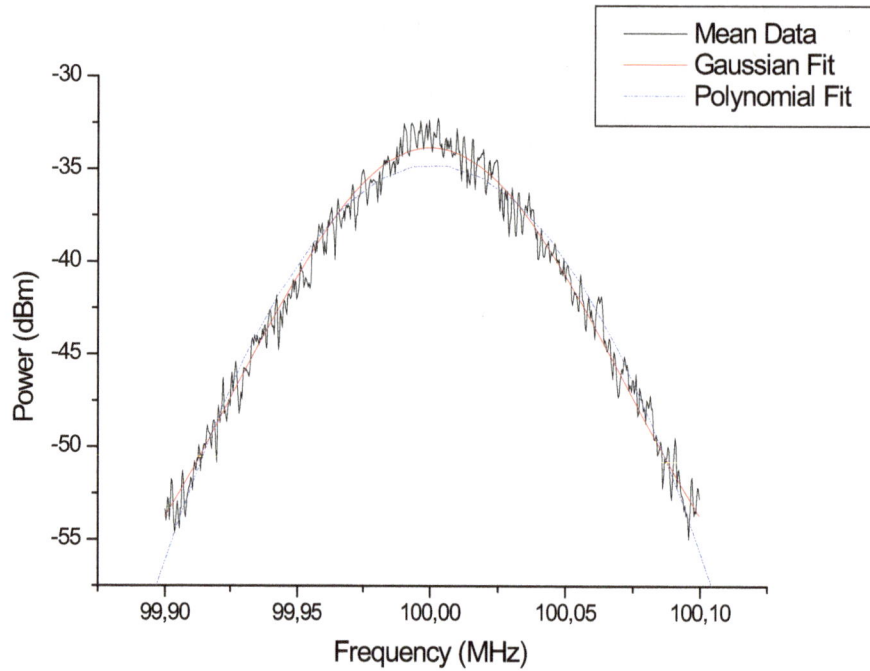

Figure 2: Mean value of a FM station and its spectrum approximation

Table 1: Gaussian fit parameters

Parameters	Values
y_o	-94.20817 dBm
A	58.92433 dBm
f_m	100 MHz
ω	0.09697 MHz
R^2	0.985

The total FM interference from all FM stations will be the sum of their PSD's

$$S_i^{'}(f) = \sum_{m=1}^{M} \hat{G}_{m(f)} = My_o + A\sum_{m=1}^{M} \exp\left[\frac{\left[f - f_1 - (m-1)\Delta f_m\right]^2}{2\omega^2}\right] \tag{35}$$

where M is the number of the FM stations, y_o, A and ω are given in Table 1, f_1 is the frequency of the first FM station and Δf_m is the carrier separation between two subsequent stations.

5. Numerical Results

For the calculation of the probability of error for single SS user interfered by multiple FM stations, we applied the CLT for the estimation of the total power spectrum of all FM stations, implementing the mean value by $f_m = \dfrac{f_1 + f_2 + ... + f_M}{M}$ and standard deviation $\omega_m = M\omega$.

We assumed that all the FM stations, M, have the same standard deviation and thus power. For best results and worst case scenario, we must consider as many stations as they can fit in the FM frequency band of 20 MHz.

Suggestively, we present the results in Figure 3 to Figure 5 applying equation (20) for a typical value of processing gain, e.g. $G_p = 50 = 17\text{dB}$, and for a typical carrier separation between the FM stations, e.g. Δf_m=500kHz. The results show the probability of error versus E_b / N_o for various signal-to-interference ratios P/P$_J$, chip rate f_c, FM stations M and frequency differences Δf between the two systems. The results are also compared using the probability of error for the ideally BPSK performance $\left(P_e = \dfrac{1}{2} erfc\left(\sqrt{\dfrac{E_b}{N_o}} \right) \right)$.

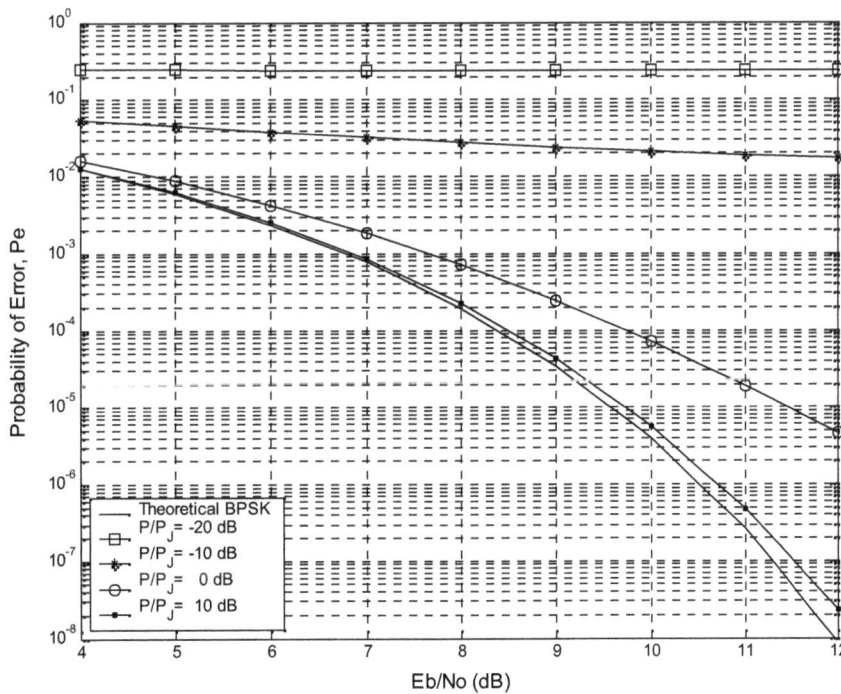

Figure 3: Bit error probability for Δf=0, f_c=10Mbps, M=40 and W_i=20MHz

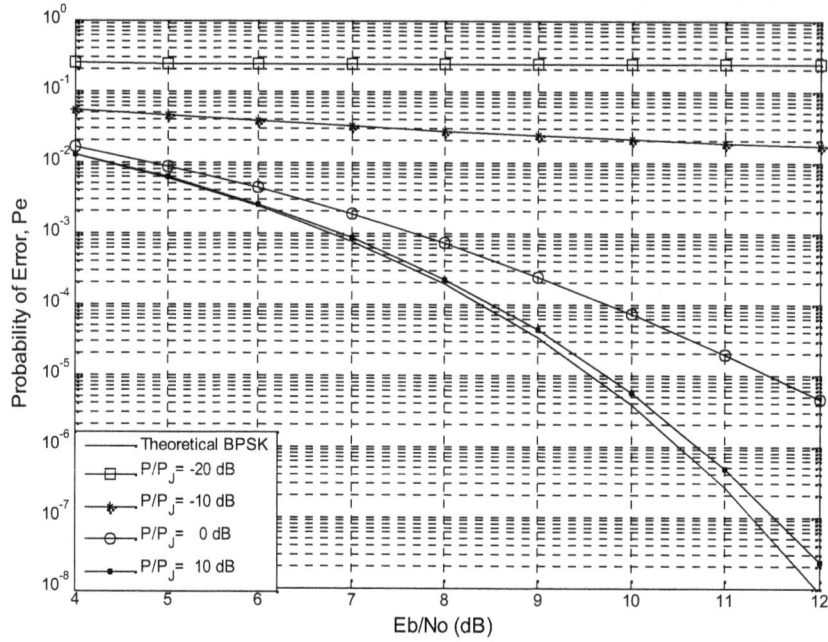

Figure 4: Bit error probability for $\Delta f = f_c/2$, $f_c = 1$Mbps, M=40 and $W_i = 20$MHz

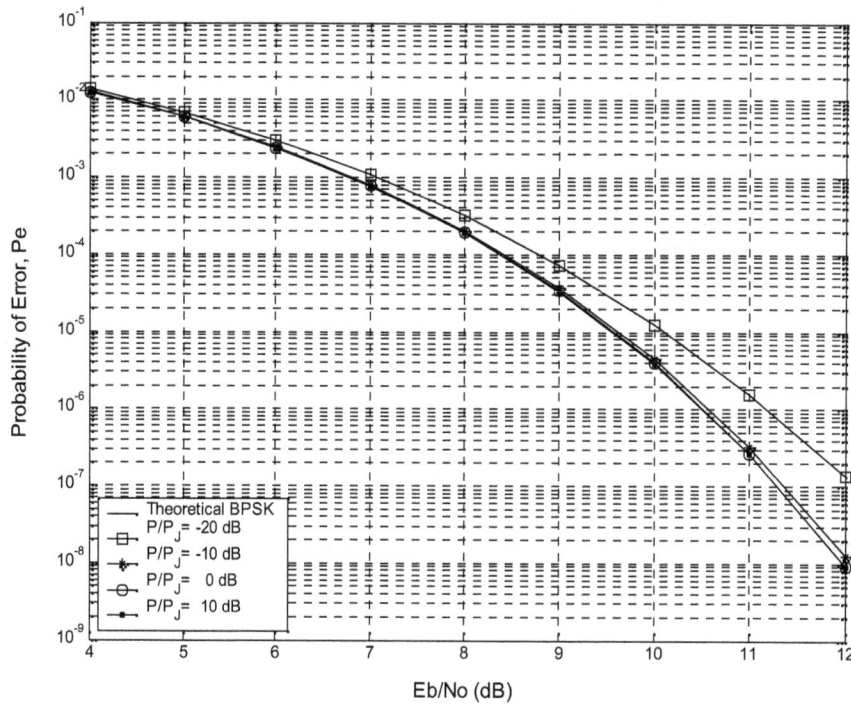

Figure 5: Bit error probability for $\Delta f = f_c$, $f_c = 5$Mbps, M=10 and $W_i = 5$MHz

As can be shown from Figures 3-5, there is dependency between the carrier spacing of these two systems, the chip rate of the SS system and the number of FM stations in the frequency band. The worst case performance of the SS system is occurring when the two systems has identical center frequency (Δf=0) and the number of FM stations are large. While the frequency offset Δf between the center frequencies of these two systems is kept as large as possible and furthermore the chip rate of the SS system has maximum values, in that case the system performs ideally, even for great extent of interference.

The performance of the SS system utilizing multiple users is derived by using equations (32) and (33). In Figure 6, the bit error probability as a function of active Spread Spectrum users is presented, for *M=40* interfering FM stations, E_b/N_o =5dB and 10dB, processing gain $G_p = 50 = 17$dB , chip rate f_c =10Mbps, frequency offset between the two systems Δf=*0* and signal-to-interference ratios P/P_J= -20dB, -10dB and 0dB. Accordingly, Figure 7 depicts the bit error probability as a function of active Spread Spectrum users, for *M=10* interfering FM stations, E_b/N_o =5dB and 10dB, processing gain $G_p = 50 = 17$dB , chip rate f_c =10Mbps, frequency offset between the two systems Δf=*5MHz* and signal-to-interference ratios P/P_J= -20dB, -10dB and 0dB. In Figure 8 the bit error probability as a function of active spread spectrum users is presented, for *M=10* interfering FM stations, E_b/N_o =5dB and 10dB, processing gain $G_p = 50 = 17$dB , chip rate f_c =10Mbps, frequency offset between the two systems $\Delta f= f_c/2$ and signal-to-interference ratios P/P_J= -20dB, -10dB and 0dB.

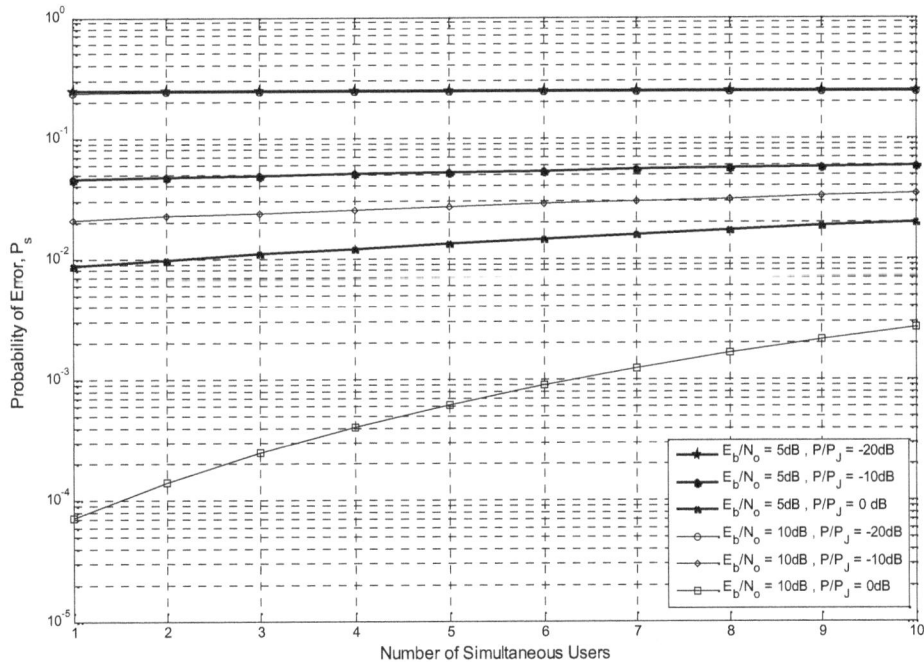

Figure 6: Bit error probability for Δf=0, f_c =10Mbps, M=40 and G_p=50

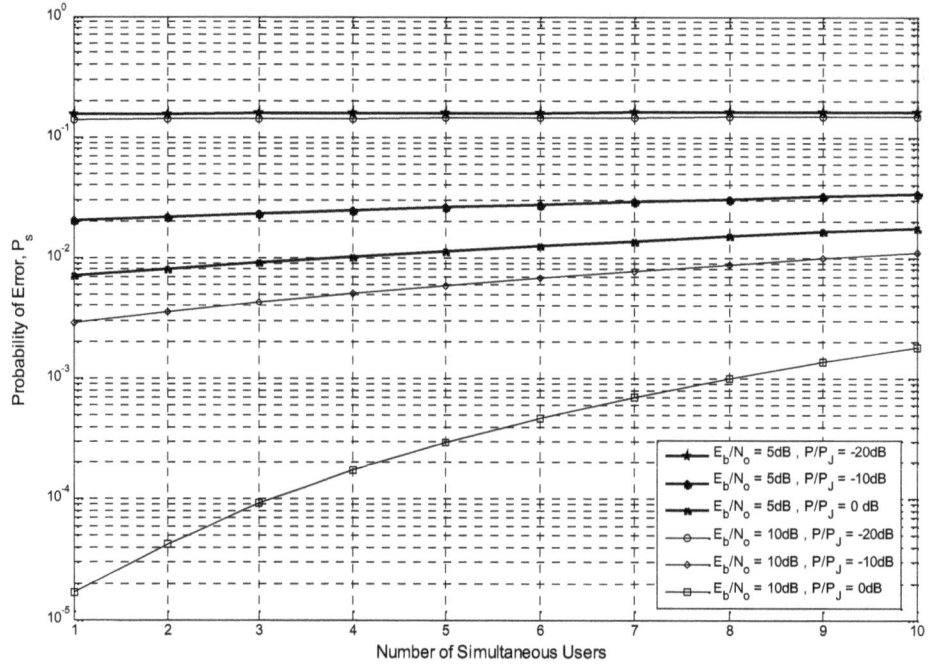

Figure 7: Bit error probability for Δf=5MHz, f_c =10Mbps, M=10 and G_p=50

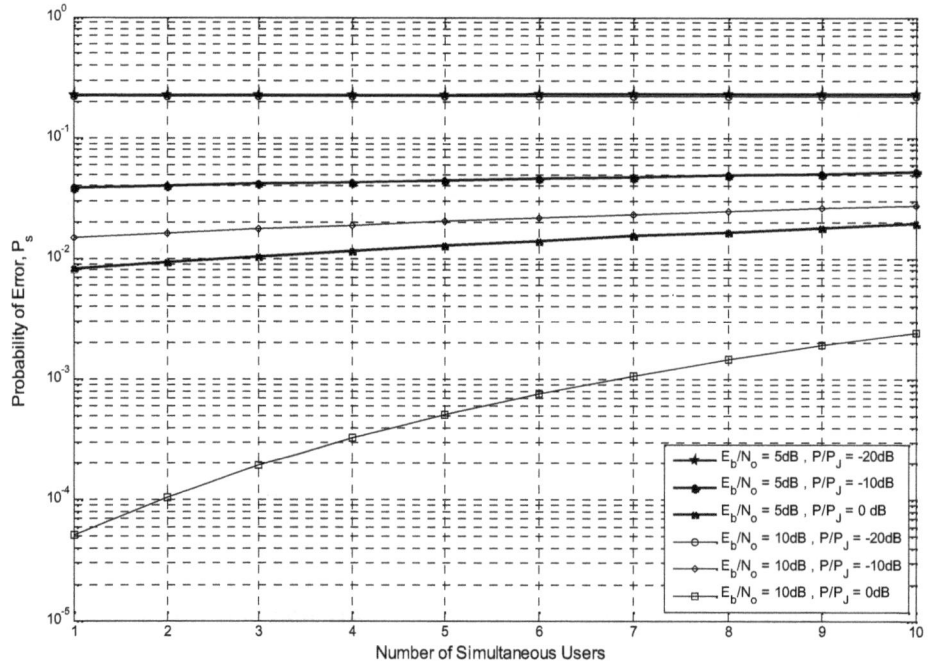

Figure 8: Bit error probability for Δf= f_c /2, f_c =10Mbps, M=10 and G_p=50

The results showed that there is a huge dependence between the spreading codes, the frequency distance between the two systems and the amount of interference, while the probability of error deteriorates when the number of simultaneous users is considerably increased.

6. Simulation Results

The reported objectives of this study will be supported with the help of an advanced design simulation system (Agilent-ADS) and evaluated in a realistic environment. The approach is based initially on theoretical level and then on simulation level. During the simulation, the effect of all factors that have analytical expression is verified, as well as these that do not have, trying to evaluate valid situations and systems [13].

The rationale of the simulation system consists of several steps. The first step comprises of the simulation of the spread spectrum transceiver, where the configuration for BPSK modulation is taken into consideration. Specifically, for the receiver's design, the correlator model of receiver has been noticed, as already discussed in Section 2 and presented graphically in Figure 2. A calibration of the system has been performed, in order to ensure for the accuracy of the process in terms of theoretical SNR vs. E_b/N_o.

The second step is the simulation of each FM station separately, so as to correspond, as much as possible, to actual conditions. Therefore, it has to be taken into account all the features that completely characterize the emitted frequency modulated signal (pre-emphasis, deviation, etc.). Likewise, the total FM spectrum is reproduced upon the recommendations of ITU for FM transmission [19], where the spectral output appears like the measured one of Figure 2. Subsequently, this module of FM station is characterized as independent subunit and the RF sum of all subunits provides the total multiple narrowband interference in the proposed system.

During the simulation process of signal transmission of SS system, we dealt with the total radio path, which is characterized by the multipath transmission, the propagation environment, the movement of receiver and the heights of the antennas. A typical propagation model for urban environment was selected, because the propagation channel effect wasn't the focal mission of this study. The proposed model for the radio path channel was incorporated into the process so as the final system of transmitter and receiver to be calibrated, without the FM interference, but with the addition of white noise, leading to a produced bit error rate similar to the theoretical ones.

Integrating all FM subunits constitutes the final step of the simulation configuration, which comprises the total model of FM stations in the final system of transmitter and receiver, illustrating the multiple narrowband interference to the system. Several number of interfered narrowband FM stations are considered during the simulation procedure.

The total performance of the SS system is measured through the estimation of the error rate at the output of the SS receiver. In order the bit error rate measurements to become more precise and efficient, the model of SS system should be regulated so as to include all the effects of their individual elements, apart from the receiver's noise. The control of system noise will become from a controlled noise source. This is what regulates also the quantity E_b/N_o.

The accuracy of BER measurements is actually only estimation and not the precise measurement, which their accuracy depends on the quantity of received samples. It is also reported as Monte Carlo approach. Generally, the BER estimation is usually very time-consuming process for practical applications. Typically, the technique of Importance Sampling [20],[21] is used, in order to substantially decrease the number of required samples. According to this technique during the simulation, the characteristics of the system are differentiated so as to vary the probability density function. Eventually, it is the preferable technique for the enhancement of BER measurements.

For the implementation of the simulation system certain constant parameters of the units and subunits have been used, in conjunction with variables ones. Because of the complexity and the large calculating time of the simulation, we varied only some of the parameters of the SS and FM system. At Table 2, the constant and the variable parameters of the simulation system are depicted. Prior to the FM interference employment, a calibration of the system has been performed, in order to ensure for the accurate feasibility of the process (theoretical SNR vs. E_b/N_o).

Table 2: Simulation parameters

Constant	*Variable*
Processing Gain (G=17dB)	SS Power
SS Bandwidth 20MHz (f_c=10Mbps)	FM Power (each station)
Frequency separation between two consequently FM carriers (500kHz)	E_b / N_o
Signal-to-Noise Ratio for SS system and each FM station (SNR = 10dB)	Variance of the error estimator
Propagation channel and antenna gain	Frequency separation (Δf) between the SS system and the central frequency of FM band

Figure 9 and Figure 10 depict the results of the simulation regarding the bit error rate at the output of the correlator for various ratios of wanted – to – unwanted signals (CIR) and frequency separation (Δf) between the SS system and the middle frequency of the FM band. At this point, firstly we must mention that the calculation of the power of the interference includes the power from all the FM stations that are contained in the predetermined by the ITU frequency band and secondly, the variance of the error estimator had the value of 0.01. The specific value was satisfactory enough, but particularly aggravating for the simulation system. Because of the time-consuming simulation process, we tried to depict results comparable to the theoretical ones, like i.e. the numerical results of the analytical expressions in Section 5.

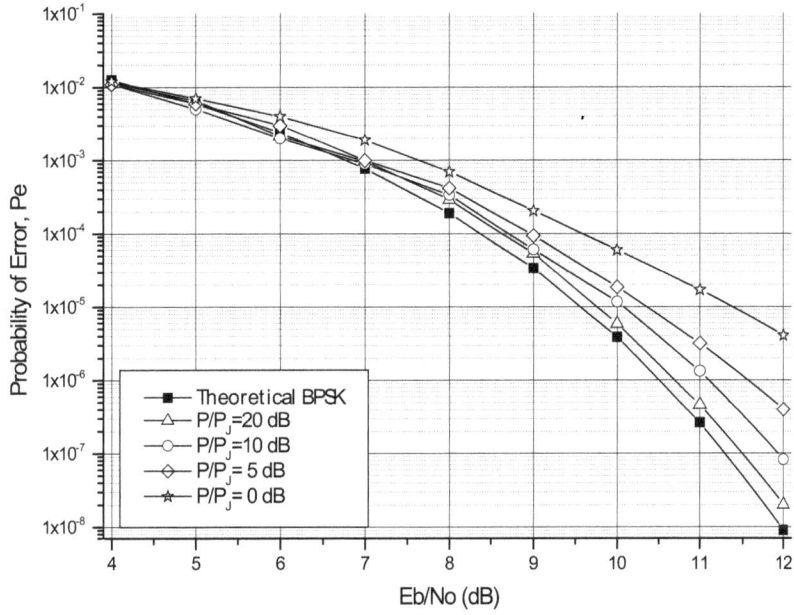

Figure 9: Bit error probability for Δf=0MHz, f_c=10Mbps, M=10 and Δf_m=500kHz

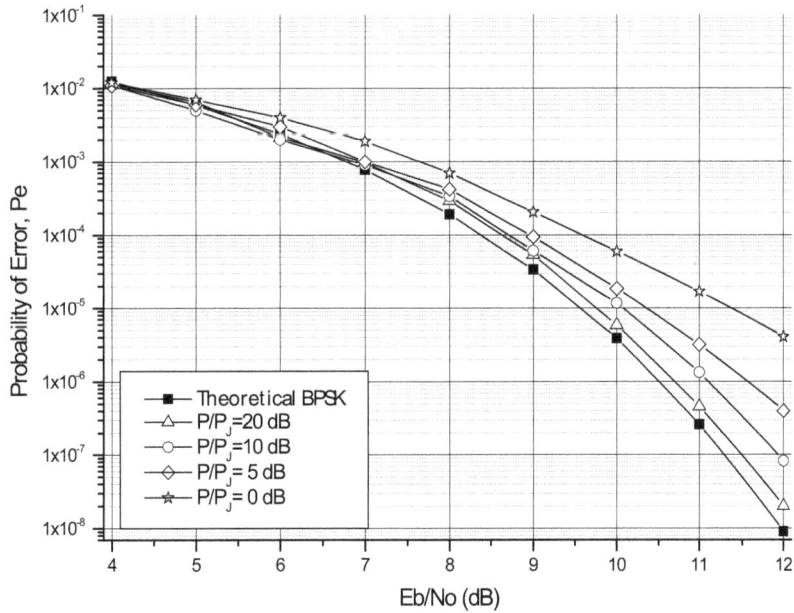

Figure 10: Bit error probability for $\Delta f= f_c /2$, f_c=10Mbps, M=10 and Δf_m=500kHz

7. Discussion and Conclusions

In this study, an efficient spectrum sharing approach has been proposed increasing the usage of real existing regulated spectrum by applying a wideband spread spectrum system. In particular, the investigation of spectrum overlay of a Direct Sequence Spread Spectrum system on the existing narrowband FM broadcasting system has been presented. We considered the implementation of a wideband SS system consisting of single and multiple users, situated on top of an existing conventional multiple narrowband FM system utilizing the 87.5 – 108 MHz band, causing interference on the SS system.

We examined the amount of multiple analog FM interference, which would not cause excessive degradation in the co-located SS system, through extensive interference analysis and simulation results. The conducting research was based on simulation of communication systems; it is critical to validate the system under development to assure that the results are indicative of the systems operation in a usable environment. The parameters of these two systems that had to be taken into consideration were several enough so as to designate the justification of this study. No channel coding has been used because our primary endeavour was to ensure cooperation between these two systems. The results of this study provide all the preliminary - but necessary - limitations attempting to spectrally coexist two dissimilar systems.

The evaluation of error probability based on simulation procedure provides interesting results that are, in some cases, close to the results obtained from the theoretical approximation in section 4, and when Figures 9 and 10 with Figures 3 and 4 are compared respectively. This implies that the assumptions and approximations proposed in section 3 and 4 and introduced in analytical expressions for the probability of error, are acceptable and barely affect the obtained results. Divergences between theoretical and simulation results are observed for high values of signal-to-interference ratio P/P_J, where the simulation procedure gives more strict interference levels.

Following the Central Limit Theorem, the Gaussian approximation for the derivation of the total FM interference and the simplified expression of the Improved Gaussian Approximation (SIGA) for multiple SS users, we conclude to the following results:

- The worst case for the probability of error occurs when the two systems have identically frequencies.
- The frequency difference Δf has minimal impact to the performance of the system, while the number of FM stations is sufficiently large.
- As the total number of the FM stations is kept large considering small frequency separation Δf_m, the curves of bit-error-rate are improved.
- The system performance is highly degraded, especially when the number of active simultaneous users is increasingly rapidly.
- For more than roughly ten users the degradation of the performance reaches a threshold over which additional users have no impact on the system performance for the same FM interfering conditions.

In general, this study gives a key motivation for the exploitation of a different radio frequency band for the physical layer, than the ones the standards provide, in wireless personal and sensors networks. Extended research activities are currently undertaken by the authors in the area of a detailed SS system optimisation, in terms of coding and interference cancellation techniques, in order to establish the maximum allowable bit rate for the data signal. However, the results depend also on the choice of broadcasting parameters, such as the acceptable interference level in the demodulated analog FM audio signal (SNR degradation), as presented in [11].

8. References

[1]   Wang, J., Ghosh, M. & Challapali K. (2011). Emerging Cognitive Radio Applications: A Survey, *IEEE Commun. Magazine*, Vol. 49, No. 3, pp.74-81.

[2]   Chen, K.-C. & Prasad, R. (2009). *Cognitive Radio Networks*, John Wiley & Sons Ltd, UK.

[3]   Milstein, L. B. *et al.* (1992). On the feasibility of a CDMA overlay for personal communications networks. *IEEE Trans. Commun.*, Vol. 10, No. 4, pp.665-668.

[4]   Hmimy, H. H. & Gupta, S. C. (1996). Overlay of cellular CDMA on AMPS forward and reverse link analysis. *IEEE Trans. Veh. Technol.*, Vol. 45, No. 1, pp.51–56.

[5]   Pickholtz, R. L., Milstein, L. B. & Schilling, D. L. (1991). Spread spectrum for mobile communications. *IEEE Trans. Veh. Technol.*, Vol. 40, No. 2, pp.313–322.

[6]   Chiani, M. & Giorgetti, A. (2009). Coexistence Between UWB and Narrow-Band Wireless Communication Systems, *Proceedings of the IEEE*, Vol. 97, No. 2, pp.231-254.

[7]   Papadopoulos, H. C. & Sundberg, C.-E. W. (1998). Simultaneous broadcasting of analog FM and digital audio signals by means of adaptive precanceling techniques. *IEEE Trans. Commun.*, Vol. 46, No. 9, pp.1233–1242.

[8]   Chen, B. & Sundberg, C.-E. W. (2000). Digital audio broadcasting in the FM band by means of contiguous band insertion and precanceling techniques. *IEEE Trans. Commun.*, Vol. 48, No. 10, pp.1634–1637.

[9]   Rome,r K. & Mattern, F. (2004). The Design Space of Wireless Sensor Networks. *IEEE. Wireless Commun.*, Vol. 11, No. 6, pp.54-61.

[10] Akyildiz, I. F., Su, W., Sankarasubramaniam, Y. & Cayirci, E. (2002). A Survey on Sensor Networks. *IEEE Commun. Magazine*, Vol. 40, No. 8, pp.102-116.

[11] Vouyioukas, D. & Constantinou, P. (2003). Performance Degradation of Analog FM System Due to Spread Spectrum Overlay. *IEEE Trans. Broadcast.*, Vol. 49, No. 2, pp.113-123.

[12] Pickholtz, R. L., Schilling, D. L. & Milstein, L. B. (1982). Theory of Spread-Spectrum Communications - A Tutorial. *IEEE Trans. Commun.*, Vol. 30, No. 5, pp.855-884.

[13] Vouyioukas, D. (2003) Interference Criteria for Spectral Overlay Systems between Direct Sequence Spread Spectrum and Analog FM Signals. Dissertation, National Technical University of Athens.

[14] Rappaport, T. S. (2002). *Wireless Communications - Principles and Practice, 2nd ed.*, Upper Saddle River: Prentice-Hall Inc.

[15] Shanmugan, K. S. & Breipohl, A. M. (1998). *Random Signals - Detection Estimation and Data Analysis*, New York: John Wiley & Sons.

[16] Morrow, K. R. & Lehrert, J. S. (1989). Bit-to-bit error dependence in slotted DS/SSMA packet systems with random signatures sequences. *IEEE Trans. Commun.*, Vol. 37, No. 10, pp.1052–1061.

[17] Holtzman, J. M. (1992). A simple accurate method to calculate spread spectrum multiple access error probabilities. *IEEE Trans. Commun.*, Vol. 40, No. 3, pp.461–464.

[18] Morrow, K. R. (1998). Accurate CDMA BER calculations with low computational complexity. *IEEE Trans. Commun.*, Vol. 46, No. 11, pp.1413-1417.

[19] ITU-R BS.450-3 (2001), *Transmission standards for FM sound broadcasting at VHF*, ITU Recommendation Standards.

[20] Lu, D. & Yao, K. (1998). Improved importance sampling technique for efficient simulation of digital communication systems. *IEEE J. Select. Areas on Commun.*, Vol. 6, No. 1, pp.67-75.

[21] Chen, J. Lu, D. Sadowsky, J. S. & Yao, K. (1993). On Importance Sampling in Digital Commuications – Part I: Fundamentals. *IEEE J. Select. Areas on Commun.*, Vol. 11, No. 3, pp.289-299.

6

IMPLEMENTING A GREEDY CHAIN ROUTING TECHNIQUE WITH SPREAD SPECTRUM ON GRID-BASED WSNS

Hossein Sharifi Noghabi [1], Arash Ghazi askar [1], Arash Boustani[2] and Arash Moghani[1], Motahareh Bahrami Zanjani[2]

[1]Department of Computer Engineering and Information Technology,
Sadjad Institute of Higher Education, IRAN
[2]Department of electrical Engineering, Wichita State University, USA
(H.sharifi219/ a.ghazi2003/ a.moghani206)@sadjad.ac.ir, (axboustani/
mxbaharamizanjani@wichita.edu)

ABSTRACT

Wireless Sensor Networks (WSN) are set of energy-limited sensors, which recently have been point of interest due to their vast applications. One of the efficient ways to consume energy in these networks is to utilize optimal routing protocols. In this approach, we proposed a greedy hierarchical chain-based routing method, named, PGC (stands for Persian Greedy Chain) which route the network applying Spread Spectrum codes as a mask given to the grid cells. Due to similarities between the proposed method in this article and LEACH protocol, we compare this routing protocol with the proposed model from diverse aspects in the simulation section such as remaining energy and being fault tolerant and reliable. The results prove that presented method is more robust and efficient.

KEYWORDS

Chain Structure, Greedy Routing, Spread Spectrum, Grid-Based Wireless Sensor Networks

1. INTRODUCTION

In recent years, Wireless Sensor Networks had been developed in many fields like medical science, military and even commercial aspects.one of the critical concerns of scientists in WSN, is optimal consuming of energy. Since sensor's batteries are not rechargeable or charging them is too costly, we should apply some special methods to minimize the energy consumption .Obviously, with reducing the energy consumption, the lifetime of the network will be maximized.

One of the solutions is to utilize an optimal routing in WSN. From structure designing view, routing techniques are divided into three parts: Flat, Hierarchical and Location-based.

In flat routing protocols, all of the nodes play same role in aggregating and sending data and routing process. In location-based routing algorithms, sensors are controlled by places they hold. In hierarchical routing algorithms nodes are hierarchical structured and network is divided into particular regions called Cluster [1].

For hierarchical structures, LEACH algorithm is discussed [2]. LEACH includes a distributed arrangement of clusters in which the Cluster Head (CH) is responsible for data aggregation and data transmission to the sink.

Another solution which leads to reducing energy consumption and consequently increasing the network lifetime in WSN, is taking advantage of Grid-cells.

In [3], it is illustrated that applying grid cells in WSN significantly reduces energy consumption. In each grid only one Gateway (GW) node should be active and sense the environment, hence, all other nodes will be slept. When the energy if the GW reaches to a minimum threshold, the GW is able to substitute another node, so the explained process will be repeated all over again. By utilizing GAF protocol; it is highly likely to save 40 - 60% of energy.

In [4] a new and secure method for data aggregation is introduced. This method is based on Grid cells, and one single mask is considered for a group of sensors, using PCC (Persian Chip Code).

This manner causes each group of sensors to be addressed uniquely and data to be delivered would be encoded. It is clear that encoding each data leads to more confidentiality and security in the network [4, 5].

Grid shape is also effective in optimizing the energy consumption of whole WSN. Results had shown that hexagonal grids due to full coverage, will improve network functionality in consuming energy, end - to - end delay, fault tolerance and security [6]. In the other types of networks dividing the environment will also increase the utilization [7].

2. PREVIOUS WORKS

2.1. Hierarchical algorithms in organizing network structure

LEACH algorithm is a cluster-based algorithm and clustering the nodes is based on receiving signal strength. The responsibility of transmitting data to the sink is with cluster heads.

In LEACH protocol, only 5 % of all nodes take the chance of being cluster head. All data processing, like data aggregation and propagation, is carried out locally in every cluster. LEACH algorithm has its own disadvantages too. Each time the algorithm is run in the network, the lifetime of sensors in the whole network will be decreased. Moreover, the energy of CH nodes is consumed more quickly than the other nodes [8]. In LEACH, data is transmitted directly from nodes to cluster heads, as a result, if there is a long distance between one of the nodes and the cluster head, the non-cluster head node should consume too much energy for delivering the information to the cluster head. This problem is strongly against LEACH, because it consumes a great deal of energy to transmit data to the sink. Nevertheless, the transmitting data might be corrupted or lost in the path and make the energy consumption much higher. LEACH also has a large overhead in producing tree among the nodes [1].

2.2. Grid-cell production algorithms

As mentioned earlier, one node can be active in each grid and alter other nodes to sleep mode. The activated node is often called Coordinator or Gateway node.GAF and SPAN are two methods for creating grids.

In GAF, the environment is divided into virtual cells and determinations about which node is sleep and which node is GW are made based on system information. The only problem in this method is that geographical locations of sensors are needed. In SPAN, each node can determine whether being coordinator or not. Coordinator election is performed periodically; just to prevent high energy consumption in GW nodes. One of the worst disadvantages of applying SPAN is electing too much coordinator, whereas in this method the total number of coordinator should be minimal [3].

2.3. Persian Chip Code

The proposed method in PCC, use a recursive algorithm and produces Orthogonal Chip Codes on a matrix in order to work on the CDMA-based networks which have the Auto Correlation, Cross Correlation and Hamming Property features. The below matrix is used for creating PCC:

$$
P_{4^n} = \begin{bmatrix} P_{4^{n-1}} & P_{4^{n-1}} & P_{4^{n-1}} & \sim\mathbf{P_{4^{n-1}}} \\ \sim P_{4^{n-1}} & \mathbf{P_{4^{n-1}}} & \sim P_{4^{n-1}} & \sim P_{4^{n-1}} \\ P_{4^{n-1}} & P_{4^{n-1}} & \sim\mathbf{P_{4^{n-1}}} & P_{4^{n-1}} \\ \mathbf{P_{4^{n-1}}} & \sim P_{4^{n-1}} & \sim P_{4^{n-1}} & \sim P_{4^{n-1}} \end{bmatrix}
$$
$$
\forall\, n \in \mathbb{N} \quad ; \quad P_{4^0} = [0]
$$

2.4. Energy aware and highly Secured Data Aggregation for Grid-based Asynchronous Wireless Sensor Networks

One of the related methods is to encode and aggregate data using PCC (Persian Chip Code). For data aggregation, the cluster heads transmit the encrypted data to the sink and the sink will decrypt it in order.at first, the sink divides the network into grid-cells, then it specify a set of PCC codes to the sensors within each grid for encoding data.as a result, the network will be addressed just like a non-WSN network with IP and Subnet [4].

3. PROPOSED MODEL

In the proposed model (PGC), a method for optimal consuming of energy and prolonging the lifetime of WSN is provided. This method will improve energy consumption in the network by making the best use of a greedy routing algorithm and smart movements of sinks in the region of interest. In this section we will explain different phases of proposed model.

3.1. Proposed Routing algorithm

After forming our hexagonal grid cells and defining the active node in each grid, the discussed algorithm in 2.4 Section will be run on all grids. After running PCC on whole network, a hierarchical address similar to IP is assigned to all of the GWs and each GW, according to its mask, will communicate with other GCs.

At first, Sink broadcasts an initial HELLO message. Then, all of the receiving messages are examined, and the nodes which their message were received with maximum transmission power will be selected as "immediate GWs". These "immediate GWs" are the nearest GWs to the sink .We introduce a parameter, namely "Next_Hop_Sink" (NHS). This parameter is given to the nearest adjacent GWs which replied the initial message with their minimum transmission power (in the other word, sink received with maximum transmission power). GWs holding NHS parameter , transmit the data directly , immediately and without any routing to the sink .Main idea in this model is that each GW transmit data to the adjacent GW with least Mask number - which was produced and assigned to each grid earlier, by PCC. This numbering method indicates that other GCs (which are non-immediate) which are nearer to the sink, would obtain the smaller mask number in comparison to the farther GCs. for example if the sink is located at the right side of the network, then the GCs located at the right side of the GCs containing "immediate" GWs, have smaller mask number. Note that similar to ESTOC [4], in this method the sink eventually has all of the data from all of the sensors in the network.

In addition to every sensor node, each GC has its own energy parameter (denoted by G for Green, Y for Yellow and R for Red) for whole GC. Green GCs are grid cells which have at least one Green node within. While all of the nodes in the GC become yellow, the GC state becomes Yellow too. It is obvious that if all the nodes become red, the GC state turns to Red. Red GCs would not participate in routing process. The priority is with Green GCs. in the absence of Green GCs in the predefined route (which were calculated earlier e

The last important thing to note is that if a GW attempts to send data and due to energy consumption, there is no way reachable to sink by its neighbor grid cells, the GW would send data directly to sink. Of course, if any available GC is present between the dead GCs and sink, data would be delivered to those GCs first.

3.2. Coding and decoding manner in proposed model

In chain-based routing, every time the data is encoded and decoded, large overhead will be imposed to the network. The solution is that sink broadcasts all the keys, at the very first moment to all of the grids. By doing this, each GW stores mapping of all grids to their corresponding keys, so encoding and decoding will be done easily. After changing the GW and electing a new one, the grid keys will be transferred from old GW to the new one.

3.3. Distributing energy consumption by moving sinks and sliding windows

To achieve the goal of distributing energy consumption, multiple sinks are used asynchronously. Whenever the average energy of GWs adjacent to the sink reach to a particular threshold, in order to prevent centralized energy consumption, another sink can be activated. It is clear that the addressing process should be start again and new "immediate" GWs will be chosen.

Now suppose that all of the sinks had been selected one time and it's the first sink's turn .if the old low-energy GWs continue being as GW, and the other GWs never have a chance to act as the "immediate" GWs, the overall lifetime of the network would be decreased. thus in proposed method , when the energy of GWs in the window reach to a particular threshold , the GWs send a message to sink and sink would consequently omit the mask address of the low-energy GW. Then sink would move to the other side of the grid and the process of choosing the "immediate" GWs would be started again. There's no need to mention that one sink should be active at a time.

3.4. Examining an instance

In order to clarify the proposed model, an instance illustrated to plot the routing technique. As it is shown in Figure.1, the proposed routing technique implies that routing through the network from source to sink should be performed in a greedy manner. First step is distributing codes generated by PCC to all grid-cells. In the example, the chip codes simplified to simple decimal numbers, thus, each grid now has a specific unique ID. Assume that GW in grid-cell 25 attempts to transmit data. The GW would send data to its least-numbered adjacent GW (here, active GW of grid-cell 19). So data would be delivered similarly to the next GW, until it reaches the grids holding NHS parameter. In this phase data would be delivered directly to the sink, located in top right of the network.

After some time which the average energy of our three grid-cells holding NHS parameter reaches to a particular threshold, the sink would move to the left side and all the process start all over again.

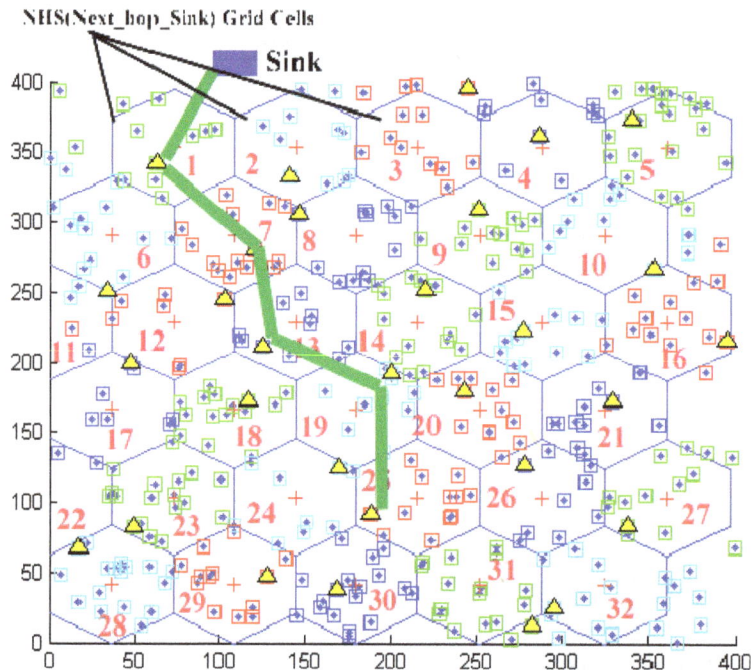

Figure 1. Routing data sourced from GC 26 to sink, passing through GCs 19 – 13 – 7 – 1 .The sliding window contains the grid numbers which hold NHS parameter.

4. Network Model

According to previous section, our proposed approach has two main advantage in data gathering in comparison to LEACH : first , in the case that data transmission is not back-to-back stream of data and data is transmitted out- of-order and with an exponential distribution , our proposed model has better performance. In periodic transmission by all of the nodes, LEACH can obtain similar performance to the proposed method, while in other case; the proposed method has better performance. In order to prove the results, extensive simulations performed with MATHLAB 2008.

The second advantage of the proposed method is that due to using PCC in proposed method, data transmission in noisy environments is much better than LEACH, and data corruption has much lower affection to the network which uses PGC, in comparison to LEACH. This will cause reducing retransmissions in network which will leads to more energy saving of nodes and increase lifetime of entire network. Remaining energy of sensors in proposed model and LEACH per 35000 packets in network is illustrated in Figure 1. Due to Figure 1, in equal times of algorithm execution, proposed model has consumed less energy than LEACH. As depicted in Figure 2, available duty cycle of our proposed model and LEACH, each of them in two cases of 99.0 % and 99.5 % packet loss is plotted. Due to using PCC in our proposed model, fault tolerance in our method is better. In each two cases, even by applying noise, receiver can decode information and obtain data.

Table 1 illustrates required information for simulation and amount of environment changes in different cases of network operations.

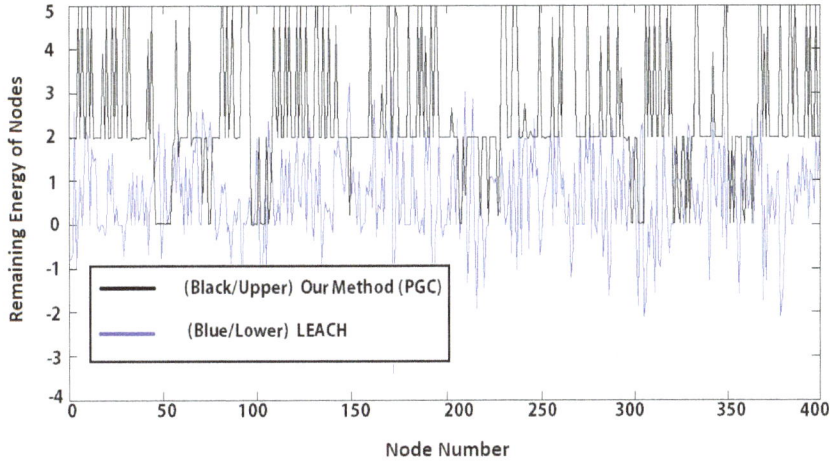

Figure 2. Remaining Energy of nodes after transmitting 35000 packets. The upper wave (depicted in black) which is our method (PGC) has more energy than nodes running LEACH protocol (which is depicted in blue, lower wave).

Table 1. Main Simulation Parameters

Parameter	Value
Network size	400*400 meter
TX Power / RX Power	15mW/10m / 13 mW/10m
Processing Power	3 mW/10m
Initial Energy per Node	5J
Minimum threshold of energy	0.03 J
Number of Nodes	400 sensors
ε fs □	10 PJ/bit/m4
Carrier frequency	2.4 - 2.48 Ghz (Zigbee)
Average of sensor per grid-cell	13 sensors
Number of grid-cell	30 grid-cell
Grid-cell structure	HCn6
Virtual grid Creator Algorithm	GAF Algorithm
Chip Code Size	32 bit- Orthogonal Asynchronous- Chip Code
Orthogonal Chip Code creator	Persian Algorithm Chip Code
Communication Multiplexing	CDMA - FHSS
Noise value	99.0 & 99.5 % packet loss

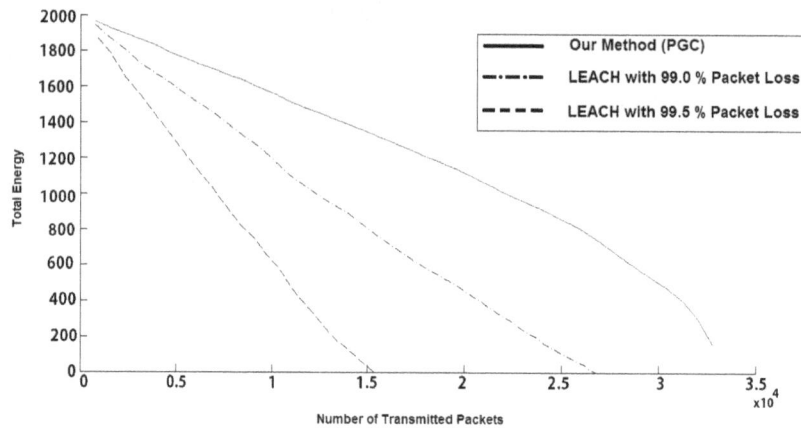

Figure 3. Evaluating two protocols in noisy environment. The upper line is our proposed model which has least packet loss per 35000 packets, the middle dotted-dashed line is LEACH with 99.0 % packet loss and the lower dashed line is LEACH with 99.5 % packet loss.

5. CONCLUSION

Obviously , energy consumption is one of the important challenges in WSN and by using an appropriate routing method , energy consumption in these networks would be enhanced and lifetime of sensors and whole network will be increased.in this paper, we provided an optimal chain-based routing method in which , sensors in a grid-based WSN choose the best path to aggregate data and transmit all of the data to the sink.in the proposed method , special Orthogonal chip codes in spread spectrum are used for addressing each GC and routing the network so that each GW in each GC , select the GC with least mask number for transferring data through it .this selection will cause reducing energy consumption significantly . Additionally, existence of multiple sinks which are activated asynchronously, and a sliding window in sink's movements, will optimize the energy consumption which consequently prolong lifetime of sensors and WSN.

6. REFERENCES

[1]   J. N. Al-Karak and A. E .Kamal, "Routing techniques in wireless sensor network: A survey" .IEEE wireless communications, Vol. 11, pp. 6-28, December 2004.

[2]   W. Heinzelman, A. Chandrakasan and H. Balakrishnan, "Energy-Efficient Communication Protocol for Wireless Microsensor Networks", Proceedings of the 33rd Hawaii International Conference on System Sciences (HICSS '00), January 2000

[3]   U. Sawant, "Grid-based Coordinated Routing in Wireless Sensor Networks", University of North Texas, Thesis Master of Science (Computer Science and Engineering), December 2006.

[4]   M. B. Zanjani, A. Boustani, "Energy aware and highly secured data aggregation for grid-base Asynchronous wireless sensor network," in proc. IEEE Pacific Rim Conference on Communications, Computers and Signal Processing, University of Victoria, Victoria, B.C., ISBN 978-1-4577-0251-8, Canada, Aug 23-26, 2011.

[5]   A. Boustani, J. Sabet, M. Azizi, N. MirMotahhary, S. Khorsandi, " 'Persian Algorithm': A new approach for generating orthogonal spreading code in spread spectrum ", IEEE Wireless Advanced, ISBN 978-1-4244-7069-3, London 2010.

[6]   S. M. Abolhasani, M. R. Meibodi , M. A. Ashari , "Developing the operation of Grid-based protocols by changing grid shape in wireless sensor networks", National conference, Amirkabir university of technology, Feb 2009.

[7]   A. Boustani, S. Khorsandi, R. Danesfahani and N. MirMotahhary, " An Efficient Frequency Reuse Scheme by Cell Sectorization in OFDMA Based Wireless Networks," in proc. IEEE, ICCIT 2009 Fourth International Conference on Computer Sciences and Convergence Information Technology, ISBN 978-0-7695-3896-9/09, Seoul 2009.

[8]   Minh-Long Pham, Daeyoung Kim , Yoonmee Doh , Seong-eun Yoo " Power Aware Chain Routing Protocol for Data Gathering in Sensor Networks" ; 14-17 Dec. 2004

7

A Fair Comparison Between Hybrid and Conventional Beamforming Receivers with Moderate Values of System Parameters

Rim Haddad[1], Ridha Bouallègue[2]

Laboratory Research in Telecommunication 6'Tel in High School of Communication of
Tunis, Route de Raoued Km 3.5, 2083 Ariana, Tunisia
[1] rim.haddad@yahoo.ca
[2] ridha.bouallegue@supcom.rnu.tn

ABSTRACT

The main role of smart antennasis to mitigate Multiple Access Interference (MAI) by beamforming (Spatial filtering) operation. In addition to MAI, the performance of receivers is limited by fast fading.
In this context, we propose in this paper a hybrid scheme of beamforming and diversity called HBF (Hierarchical Beamforming) and we propose a system model for the mathematical characterization of HBF for the performance evaluation.Moreover, we compare the performance of HBF receiver with conventional Beamforming (CBF) one. The proposed model conforms the benefits of adaptive antennas in reducing the overall interference level (intercell/intracell) and to find an accurate approximation of the error probability.

KEYWORDS

Beamforming, Hierarchical Beamforming (HBF), Conventional Beamforming (CBF), BER, Rayleigh fading

1. INTRODUCTION

As the growing demand for mobile communications is constantly increasing, the need for better coverage, improved capacity and higher transmission quality rises. Thus, a more efficient use of the radio spectrum is required. Smart antenna systems have emerged as one of the most efficiency and improving the performance of present and future wireless communication systems.

Several smart antenna systems have been proposed and demonstrated at the base station (BS) of the wireless communication system, and these have shown that significant increases in capacity are possible.

The deployment of smart antennas at cellular base station installations has gained enormous interest because it has the potential to increase cellular system capacity, extend radio coverage, and improve quality of services [1,2].

In a typical mobile environment, signals from users arrive at different angles to the base station and hence antenna arrays can be used to an advantage. Each multipath of a user may arrive at a different angle, and this angle spread can be exploited using an antenna array [3].

The Bit Error Rate (BER) is considered to be one of the most important performance measures for communication systems and hence it has been extensively studied.

In our paper, we propose a novel approach to evaluate the average probability of error by considering an approximation of the spatial filter. Hence, we will derive an analytical model for evaluating the mean BER of two smart antenna receivers: the HBF (Hierarchical Beamforming) receiver and the CBF (Conventional Beamforming) receiver.

The analysis is performed assuming Rayleigh fading multipath environments. We assume to make a comparison between HBF and CBF receivers.

We organize the rest of the paper as follows: In section 2, we introduce the system model, followed by the receiver model in section 3. The general simulation assumptions and simulation results are provided in section 4 and section 5 respectively. We conclude in section 6.

2. SYSTEM MODEL

2.1.System Modelof Hierarchical Beamforming:

We consider a BS serving a single $120°$ angular sector. It is assumed that the BS is equipped with F co-linear sub-beamforming arrays. The number of array elements in each sub-array is B. That's why the total number of array elements is $M = F \times B$. The inter-element spacing in each sub-array is $d = \lambda/2$, while the spacing between the adjacent sub-beamforming arrays (d_S), is assumed large enough $(d_S = 20\lambda$ or more) to uncorrelated fading. The extreme case of $F = 1$ and $B = M$ corresponds to the conventional Beamforming.

As the required spacing between sub-arrays for space diversity is much smaller than the sector radius, this AoA is assumed to be the same at each sub-array [4].

We consider K the total number of active Mobile Stations (MS) in the system, which are randomly distributed in the azimuthal direction, along the arc boundary of the sector cell in the far field of the array. Let's assume that $k = 1$ be the user of interest.

In this section, we consider that the BS is equipped with a hierarchical Beamforming receiver. Each sub array employs the functional block diagram of OQPSK receiver model.

2.1.1. Transmitted Signal:

We assume that the MS transmitter of each user employs offset Quadrature Phase Shift Keying (OQPSK) M-ary orthogonal modulation.

The transmitted signal s_k of the kth user can be written as [5]:

$$s_k(t) = W_k^{(q)}(t)a_k(t)a^{(I)}(t)\cos(w\omega_c t) + W_k^{(q)}(t - T_0)a_k(t - T_0)a^{(Q)}(t - T_0)\sin(\omega_c t) \tag{1}$$

where $W_k^{(q)}$ is a Hadamard-Walsh function of dimension Q which represents the qth orthogonal signal $(q = 1,2,...,Q = 64)$ of the kth user, $a^{(I)}$ and $a^{(Q)}$ are the in-phase and quadrature phase pseudo-Noise (PN) sequences, $a_k(t)$ is the kth user long code sequence, T_0 is the half chip delay for OQPSK signals, $\omega_c = 2\pi f_c$ and f_c is the carrier frequency.

2.1.2. Channel Model:

We assume in the following sections that the transmitter signal propagates over Rayleigh fading multipath channel.

The complex equivalent representation of the channel impulse response between the lth multipath of the the kth user and the the bth antenna in the fth sub-array is given as:

$$\tilde{h}_{k,l,b}^{(f)}(t) = \beta_{k,l}^{(f)} e^{-j\varphi_{k,l,b}^{(f)}} \delta(t - \bar{\tau}_{k,l}) \tag{2}$$

Where $\beta_{k,l}^{(f)}$ is the path amplitude, $\varphi_{k,l,b}^{(f)}$ is the overall path phase and $\bar{\tau}_{k,l}$ is the path delay respectively. To simplify our work, we assume that multipath channel parameters $\beta_{k,l}^{(f)}$ and $\varphi_{k,l,b}^{(f)}$ remain constant in the duration of Walsh symbol. In vector notation, the spatial signature or channel response vector $h_{k,l}^{(f)}(t)$ is given by:

$$h_{k,l}^{(f)} = \begin{bmatrix} h_{k,l,1}^{(f)} & h_{k,l,2}^{(f)} & \dots & h_{k,l,B}^{(f)} \end{bmatrix}^T \tag{3}$$

2.1.3. The received signal:

At the receiver, the total received signal for the fth sub-array can be written in vector notation as:

$$r^{(f)}(t) = \sum_{k=1}^{K} \sum_{l=1}^{L} s_k(t - \tau_{k,l}) h_{k,l}^{(f)}(t) + \eta^{(f)}(t) \tag{4}$$

where $\tau_{k,l} = \Gamma_k + \bar{\tau}_{k,l}$, Γ_k is the random delay of the kth user due to the effect of asynchronous transmission, $\eta^{(f)}$ is the noise which is assumed to be Additive White Gaussian Noise (AWGN) and $h_{k,l}^{(f)}(t)$ the channel response vector given in (2.1.2).

2.2. System Model of Conventional Beamforming:

The choice of the second model is based on the reverse link (mobile to base station) of the 3G CDMA 2000 Systems. We consider K the total number of active Mobile Stations (MS) in the system, which are randomly distributed in the azimuthal direction, along the arc boundary of the sector cell in the far field of the array. For simplicity, the conventional encoder and interleaver are ignored (this approach is widely used [1] for wireless communication systems employing multiple antennas).

2.2.1. Transmitted Signal:

The transmitted signal s_k of the kth user can be written as [3]:

$$s_k(t) = W_k^{(q)}(t) a_k^{(I)}(t) \cos(\omega_c t) + W_k^{(q)}(t - T_0) a_k^{(Q)}(t - T_0) \sin(\omega_c t) \tag{5}$$

Where $q = 1, 2, \dots, Q$, $W_k^{(q)}(t)$ is a Hadamard-Walsh function of dimension Q which represents the qth orthogonal signal of the kth user's long code sequence, $a_k(t)$ is the kth user's long code sequence, $a_k^{(I)}(t)$ and $a_k^{(Q)}(t)$ are the in-phase and quadrature phase pseudo-noise (PN) sequences, $T_0 = T/2$ is the delay for OQPSK signals.

The power of each user is assumed unity (perfect power control). To simplify our study the PN codes are presented as follows:

$$a_k^{(I)}(t) = \sum_r a_{k,r}^{(I)}(t)\, p(t - T_C) \tag{6}$$

$$a_k^{(Q)}(t) = \sum_r a_{k,r}^{(Q)}(t)\, p(t - T_C) \tag{7}$$

Where $a_{k,r}^{(I)}$ and $a_{k,r}^{(Q)}$ are i.i.d variables taking the values ± 1 with equal probability and $p(t)$ is the chip pulse shape which is assumed to be rectangular.

The equation (5) can be written as follows:

$$s_k(t) = \mathcal{R}\left\{\left[W_k^{(q)}(t)a_k^{(I)}(t) + j\,W_k^{(q)}(t - T_0)a_k^{(Q)}(t - T_0)\right]e^{-j\omega_c t}\right\} \tag{8}$$

$$s_k(t) = \mathcal{R}\{\widetilde{s_k}(t)e^{-j\omega_c t}\}$$

Where $\widetilde{s_k}(t) = S_k^{(I)}(t) + jS_k^{(Q)}(t)$ is the complex low pass equivalent of the transmitted signal.

2.2.2. Channel Model:

The kth user propagates through a multipath channel with (AoA) θ_k. We use the channel model presented in chapter 3. The complex equivalent representation of the channel impulse response between the lth multipath of the kth user and the nth element of array antenna is presented as follows:

$$\tilde{h}_{k,l,n}(t) = \beta_{k,l}\, e^{-j\left(\Phi_{k,l} + 2\pi\frac{d}{\lambda}(n-1)\sin\theta_k\right)}\delta\left(t - \bar{\tau}_{k,l}\right) \tag{9}$$

$$\tilde{h}_{k,l,n}(t) - \beta_{k,l}\, e^{-j\varphi_{k,l,n}}\delta\left(t - \bar{\tau}_{k,l}\right)$$

where $\beta_{k,l}$, $\Phi_{k,l}$ and $\bar{\tau}_{k,l}$ are the path gain, phase and delay respectively, $\varphi_{k,l,n}$ is the overall phase which includes the path phase and the difference in propagation delays between the antennas. In this case of transmitter we assume that path gains follow the Rayleigh and Ricean distributions respectively.

To simplify our work, we assume that multipath channel parameters $\beta_{k,l}(t)$ and $\varphi_{k,l,n}(t)$ remain constant in the duration of Walsh symbol [6], so $\beta_{k,l}(t) = \beta_{k,l}$ and $\varphi_{k,l,n}(t) = \varphi_{k,l,n}$ for $t\epsilon[0, T_W]$, where T_W is the Walsh symbol period.

2.2.3. Received Signal:

At the receiver, the received signal at the nth antenna element can be written as:

$$r_n(t) = \sum_{k=1}^{K} \sum_{l=1}^{L} \mathcal{R}\left\{ \left[\tilde{s}_k(t - \Gamma_k) * \tilde{h}_{k,l,n}(t) \right] e^{-j\omega_c t} \right\} + \eta(t)$$

$$
\begin{aligned}
r_n(t) = \sum_{k=1}^{K} \sum_{l=1}^{L} & \left[\beta_{k,l} W_k^{(q)}(t - \tau_{k,l}) a_k^{(I)}(t - \tau_{k,l}) \cos(\omega_c t + \varphi_{k,l,n}) \right. \\
& \left. + \beta_{k,l} W_k^{(q)}(t - T_0 - \tau_{k,l}) a_k^{(Q)}(t - T_0 - \tau_{k,l}) \sin(\omega_c t + \varphi_{k,l,n}) \right] \\
& + \eta(t)
\end{aligned}
\tag{10}
$$

where $\tau_{k,l} = \Gamma_k + \bar{\tau}_{k,l}$, and Γ_k is the random delay of the kth user due to the effect of asynchronous transmission.

3. RECEIVER MODEL:

3.1. The HBF Receiver Model:

The HBF receiver is divided in four main blocks which can be identified as follows: (1) the sub-array antenna blocks (2) the PN dispreading, (3) the Beamforming and (4) Walsh correlation and demodulation. Figure 1shows the functional block diagram of the HBF receiver.

The received signal at each sub-array antenna is first down converted. Each resolvable path is then detected by one of the RAKE fingers. To detect the lth path, the signal at the different sensors is dispread using the sequence of the respective mobile and synchronized to the delay of the lth path. The post PN-despread signal vector is:

$$Y_{k,l}^{(f)} = \left[y_{k,l,1}^{(f)} \quad y_{k,l,2}^{(f)} \quad \cdots \quad y_{k,l,B}^{(f)} \right]^{T} \tag{11}$$

In the next step, the signal after PN dispreading is combined by the Beamforming process. The Beamforming output is given by:

$$z_{k,l}^{(f)}(t) = \left(W_{k,l}^{(f)} \right)^{H} Y_{k,l}^{(f)} \tag{12}$$

Where $W_{k,l}^{(f)}$ is the Maximum SNR Beamforming weight vector given by:

$$W_{k,l}^{(f)} = \left[W_{k,l,1}^{(f)} \quad W_{k,l,2}^{(f)} \quad \cdots \quad W_{k,l,B}^{(f)} \right]^{T} \tag{13}$$

To simplify our work, we assume that the weights are set equal to the channel response vector for the desired user. This provides a lower bound on the system performance.

The last step is the correlation of the beamformers with stored replicas of the Walsh functions and then the overall decision variable is obtained by Equal Gain Combining (EGC) of all the decision variables from the multipath signals for the fth sub-array. The overall decision is then made by selecting the decision outcomes from the respective sub-beamforming array with the best channel state [4].

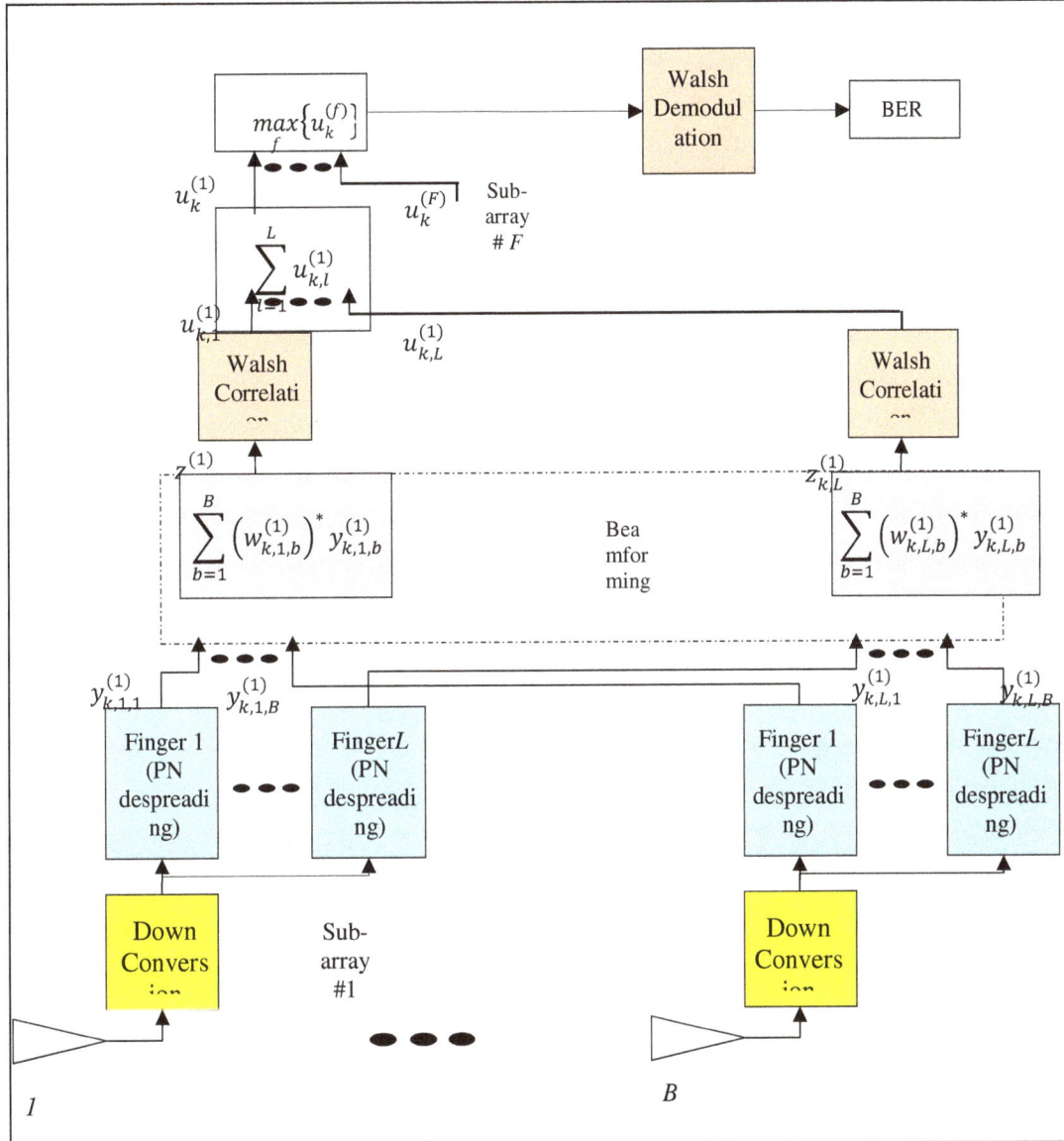

Figure 1: Receiver block diagram for Hierarchical Beamforming

3.2. The CBF Receiver Model:

The receiver is divided in four main blocks which can be identified as follows: (1) the array antenna block, (2) the PN despreading, (3) the Beamforming and (4) Walsh correlation and demodulation. We will explain the function of each block:

The first step of the receiver is to obtain the quadrature components at each antenna. We multiply the received waveforms by $\cos(\omega_c t)$ and $\sin(\omega_c t)$ respectively and then Low Pass Filtering (LPF) to remove the double frequency components that results from multiplication [7]. The output of the I-channel and Q-channel low pass filter is given by:

$$r_{k,l,n}^{(I)}(t) = \left[r_{k,l,n}(t) \cos(\omega_c t) \right]_{\text{LPF}} \tag{14}$$

$$
\begin{aligned}
&= \left\{ \beta_{k,l} W_k^{(q)}(t - \tau_{k,l}) a_k^{(I)}(t - \tau_{k,l}) \frac{\cos \varphi_{k,l,n}}{2} \right. \\
&\quad\quad \left. + \beta_{k,l} W_k^{(q)}(t - T_0 - \tau_{k,l}) a_k^{(Q)}(t - T_0 - \tau_{k,l}) \frac{\sin \varphi_{k,l,n}}{2} \right\} \\
&\quad + \eta^{(I)}(t) \\
r_{k,l,n}^{(Q)}(t) &= \left[r_{k,l,n}(t) \sin(\omega_c t) \right]_{LPF} \\
&= \left\{ \beta_{k,l} W_k^{(q)}(t - \tau_{k,l}) a_k^{(Q)}(t - T_0 - \tau_{k,l}) \frac{\cos \varphi_{k,l,n}}{2} \right. \\
&\quad\quad \left. - \beta_{k,l} W_k^{(q)}(t - T_0 - \tau_{k,l}) a_k^{(I)}(t - \tau_{k,l}) \frac{\sin \varphi_{k,l,n}}{2} \right\} \\
&\quad + \eta^{(Q)}(t)
\end{aligned}
\tag{15}
$$

The complex low pass of the received signal can be written as:

$$
\tilde{r}_{k,l,n}(t) = r_{k,l,n}^{(I)}(t) + j r_{k,l,n}^{(Q)}(t)
\tag{16}
$$

After filtering, each path is detected by one of the fingers immediately following the radio-frequency stages.

The complex low pass equivalent of the post PN-despread signal is given as yk,l,n(t) :

$$
yk,l,nt = y_{k,l,n}^{(I)}(t) + j y_{k,l,n}^{(Q)}(t)
\tag{17}
$$

The despreading sequences are denoted as [8]: $\tilde{a}(t) = a_k^{(I)}(t - \tau_{k,l}) + j a_k^{(Q)}(t - T_0 - \tau_{k,l})$
We can also write as follows:

$$
y_{k,l,n}^{(I)}(t) = \mathcal{R}\{(\tilde{a}(t), \tilde{r}_{k,l,n}(t))\} = r_{k,l,n}^{(I)}(t) a_k^{(I)}(t - \tau_{k,l}) + r_{k,l,n}^{(Q)}(t) a_k^{(Q)}(t - T_0 - \tau_{k,l})
\tag{18}
$$

$$
y_{k,l,n}^{(Q)}(t) = \mathcal{I}\{(\tilde{a}(t), \tilde{r}_{k,l,n}(t))\} = r_{k,l,n}^{(I)}(t) a_k^{(Q)}(t - T_0 - \tau_{k,l}) - r_{k,l,n}^{(Q)}(t) a_k^{(I)}(t - \tau_{k,l})
\tag{19}
$$

Where $(a, b) = a \cdot b^*$ the product between complex numbers. $\tilde{y}_{k,l,n}$ can be written in vector notation as:

$$
Y_{k,l} = \left[y_{k,l,1}, y_{k,l,2}, \dots, y_{k,l,M} \right]^T
\tag{20}
$$

In the next step, the signal after PN despreading is combined by the beamformer. In the Beamforming operation, the signals received by antenna elements are weighted by complex weights and then summed up.
The smart antenna output is given by:

$$
Z_{k,l} = \left(W_{k,l} \right)^H Y_{k,l}
\tag{21}
$$

$$
\tilde{Z}_{k,l}(t) = Z_{k,l}^{(I)}(t) + j Z_{k,l}^{(Q)}(t)
\tag{22}
$$

Where $W_{k,l}$ is the Beamforming weight vector given by:

$$
W_{k,l} = \left[W_{k,l,1}, W_{k,l,2}, \dots, W_{k,l,M} \right]^T
\tag{23}
$$

To simplify our work, we assume that the weights are set as $W_{k,l} = h_{k,l}$ and these vector channel coefficients are assumed to be perfectly known. This provides the best case system performance. The last step is the correlation of the smart antenna output with stored replicas of the Walsh functions to form the decision variable for demodulation.

The output of the qth Walsh correlator ($q = 1, 2, \dots, Q$) for single antenna is:

$$Z_{k,l}^{(I)}(q) = \frac{1}{T_W} \int_{\tau_{k,l}}^{T_W+\tau_{k,l}} \left[Z_{k,l}^{(I)} W^{(q)}(t - \tau_{k,l}) + Z_{k,l}^{(I)} W^{(q)}(t - T_0 - \tau_{k,l}) \right] dt \tag{24}$$

$$Z_{k,l}^{(Q)}(q) = \frac{1}{T_W} \int_{\tau_{k,l}}^{T_W+\tau_{k,l}} \left[Z_{k,l}^{(Q)} W^{(q)}(t - \tau_{k,l}) + Z_{k,l}^{(Q)} W^{(q)}(t - T_0 - \tau_{k,l}) \right] dt \tag{25}$$

The decision variable for the lth multipath of the kth user is obtained from the previous values:

$$u_{k,l}(q) = \left(Z_{k,l}^{(I)} \right)^2 + \left(Z_{k,l}^{(Q)} \right)^2 \tag{26}$$

The overall decision variable is obtained by Equal Gain Combining (EGC) of all the decision variables from the L multipaths as [9]:

$$u_k(q) = \sum_{l=1}^{L} u_{k,l}(q) = \sum_{l=1}^{L} \left[\left(Z_{k,l}^{(I)} \right)^2 + \left(Z_{k,l}^{(Q)} \right)^2 \right] \tag{27}$$

Finally, the receiver makes a hard decision on the qth symbol of the kth user by using the Maximum Likelihood Criteria rule as:

$$\hat{q} = \arg_{q=1,...,Q} \max\{u_k(q)\} \tag{28}$$

4. GENERAL SIMULATION ASSUMPTIONS

The performance of HBF array antenna systems is evaluated by means of Montecarlo simulations runs over the variable of interest (E_b/N_0 or M). The figure of merit used in this work is the mean Bit Error Rate (BER). This is the mean BER taken over the set of channel Rayleigh fading parameters.

The performance metric is collected and averaged over $M_c = 100$drops. A drop is defined as a simulation run for a given number of MS. During a drop, the MS's AoA increases or decreases linearly with angle change $\Delta\theta$ to crossover the entire sector azimuth range [-60°,60°]. During a drop, the channel undergoes fast fading according to the motion of the MS's. To simulate the MS mobility, we assume that the snapshot rate is equal to the Walsh symbol rate and the angle change between snapshots is $\Delta\theta = 0,01°$ per snapshot (MS travelling at 300km/h at only 100m from the BS, this value is widely used in simulations).

For clarity of investigations, the main parameters for HBF simulation assumptions are discussed below:

   a) Number of Antenna elements: To make the comparison between HBF and CBF, it is merely assumed that the number of antenna elements M is the same for both cases.
   b) Number of HBF branches: We consider in simulations that the BS is equipped with F=2 co-linear sub-beamforming arrays. This choice of sub-arrays is motivated by practical array size considerations and is relevant to a BS serving three sectors, each covering 120° in azimuth.
   c) Channel: The channel considered is Rayleigh fading with L=1,2 paths/user respectively.
   d) Pdf in AoA: We assume a Gaussian pdf in AoA. The angular distribution of the waves arriving at the BS in azimuth is described by the pdf in AoA.
   e) Angle Spread: The values of angle spread used in simulations lie in the range 5°-15° which corresponds to urban macrocellular areas.

5. SIMULATION RESULTS:

The performance of HBF is determined by the interaction of a number of factors. These include: Beamforming gain via closely spaced antenna elements within each sub-array beamforming, space diversity gain via widely separated sub-arrays beamforming, additional space diversity gain via angle spread and temporal diversity gain via the multipaths. We present in the following sections the impact of each parameter in the performance of HBF and we will make a fair comparison between HBF and CBF.

5.1. Effect of varying Noise level:

First of all, we study the performance of HBF and CBF for the case of a single user (K=1). Obviously, there is no MAI for the case of one user. We can notice from Figure 2 that both CBF and HBF for different number of antennas show a considerable improvement in mean BER compared to the conventional receiver (super imposed as reference). Besides, the improvement in mean BER increases with E_b/N_0. It is very clear from the figure that the performance of HBF is superior to CBF, e.g for a BER threshold of 10^{-2}, M=4 antennas, and E_b/N_0 of about 5dB is required for CBF, but only 2.5dB is required for HBF. The performance of HBF is superior to CBF due to space diversity gain offered by the widely separated sub-arrays, which is dominant factor (in the absence of MAI) for the case of a single user.

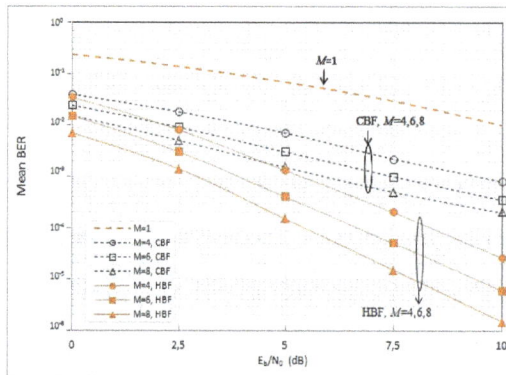

Figure2: Mean BER versus Eb/N0 for K=1 user, L=2 paths Rayleigh fading channel, σAoA=0°

5.2. Effect of varying Angle Spread:

We can notice from Figure 3 that, both CBF and HBF improve the performance as the angle spread σ_{AoA} increases from 5° to 10°. It is obvious from the figure, that for low E_b/N_0, CBF is slight better than HBF. But, as E_b/N_0 gets higher, diversity gain becomes dominant and HBF becomes better than CBF.

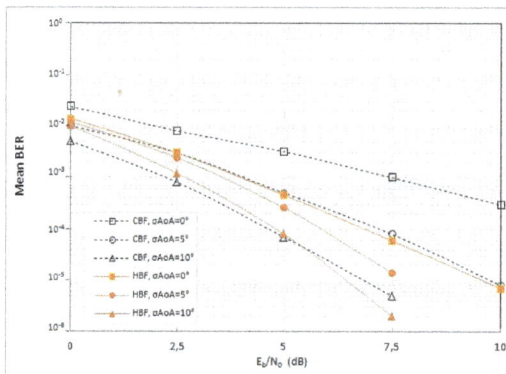

Figure3: Mean BER versus Eb/N0 for K=1 user, L=2 paths, M=6 antennas

5.3. Effect of varying Number of antennas

It is noticed from the Figure 4, that for $\sigma_{AoA} = 0°$, HBF is better than CBF due to diversity gain provided by array architecture. Moreover, there is no much improvement in performance for both CBF and HBF, by doubling the number of antennas from 4 to 8. If we want to compare angle spread scenarios, for $\sigma_{AoA} = 5°$, HBF is better than CBF, but for larger angle spreads for $\sigma_{AoA} = 10°$ and $15°$, both array architectures show a similar performance for the number of users considered in simulations.

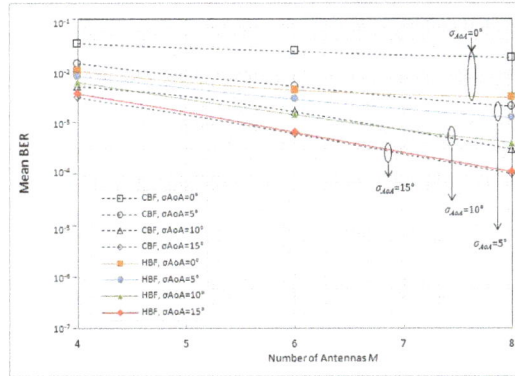

Figure4: Mean BER versus number of antennas M, K=15 users, L=1path/user

6. CONCLUSIONS:

In this paper, we have reported on the performance of hybrid scheme of diversity and Beamforming. Furthermore, its performance is compared with conventional Beamforming with moderate values of the system parameters such as angle spread number of antennas, number of multipath and number of users. It has be shown that while assuming zero angle spread, the performance of HBF is superior to CBF due to space diversity gain afforded by the well separated sub-arrays. The inclusion of angle spread produces spatial fading across the array, which results in additional diversity gain and improves the performance of both CBF and HBF schemes. For the case of moderate or large angle spread, when path diversity is present and the system is heavily loaded, CBF yields better mean BER results than HBF. All these results are based on the assumption of perfect channel estimation, that's why the choice of optimum receiver architecture is dependent on the channel conditions.

REFERENCES

[1] S. Bellofiore et al., "Smart antenna system analysis, integration and performance for Mobile Ad-Hoc Networks (MANET's)," IEEE Trans. AntennasPropagat., vol. 50, pp. 571–581, May 2002.

[2] M. Ho, G. Stuber, and M. Austin, "Performance of switched-beam smart antennas for cellular radio systems," IEEE Trans. Vehic. Technol., vol.47, pp. 10–19, Feb. 1998.

[3] R.Haddad, R.Bouallègue, "BER Performance in Space-Time Processing receiver using Adaptive Antennas over Rayleigh Fading Channels", Proc. IEEE International Conference on signal Processing and Communication , pp.1483-1486 Nov. 2007.

[4] B. A. Bjerke, Z. Zvonar, and J. G. Proakis, "Antenna diversity combining aspects for WCDMA systems in fading multipath channels", IEEE Transactions on Wireless Communications, vol. 3, no. 1, pp. 97-106, Jan. 2004.

[5] MonzigoA.Roberts, Miller Thomas, Introduction to Adaptive Arrays. Sc Tech Publishing, 2004.

[6] R.Haddad, R.Bouallègue, "BER Performance of Smart Antenna Systems Operating over Rayleigh fading Channels", proc. IEEE Wireless Days 2008, pp.1-5, Nov. 2008.

[7] T. S. Rappaport, Wireless Communications: Principles and Practice, 2nd ed. Prentice Hall, 2002.

[8] G. L. Stuber, Principles of Mobile Communication, 2nd ed. Kluwer Academic Publishers, 2001.

[9] C. D. Iskander and P. T. Mathiopoulos, "Performance of multicode DS/CDMA with M-ary orthogonal modulation in multipath fading channels," IEEE Transactions on Wireless Communications, vol. 3, no.1, pp. 209-223, Jan. 2004.

IMPACT OF CHANNEL PARTITIONING AND RELAY PLACEMENT ON RESOURCE ALLOCATION IN OFDMA CELLULAR NETWORKS

Sultan F. Meko

IU-ATC, Department of Electrical Engineering
Indian Institute of Technology Bombay, Mumbai 400 076, India
Email: sultanfeisso@ee.iitb.ac.in

ABSTRACT

Tremendous growth in the demand for wireless applications such as streaming audio/videos, Skype and video games require high data rate irrespective of user's location in the cellular network. However, the Quality of Service (QoS) of users degrades at the cell boundary. Relay enhanced multi-hop cellular network is one of the cost effective solution to improve the performance of cell edge users. Optimal deployment of Fixed Relay Nodes (FRNs) is essential to satisfy the QoS requirement of edge users. We propose new schemes for channel partitioning and FRN placement in cellular networks. Path-loss, Signal to Interference and Noise Ratio (SINR) experienced by users, and effects of shadowing have been considered. The analysis gives more emphasis on the cell-edge users (worst case scenario). The results show that these schemes achieve higher system performance in terms of spectral efficiency and also increase the user data rate at the cell edge.

KEYWORDS

FRN deployment, frequency reuse, OFDMA, multi-hop, outage probability

1. INTRODUCTION

Modern cellular networks provide various types of real-time and non real-time services. The amount of available resources (i.e., time-slot and subcarriers) and the Quality of Service (QoS) requirements determine the capacity of the cell while the transmit power and propagation conditions determine the cell size. Increased capacity, coverage and throughput are the key requirements of future cellular networks. To achieve these, one of the solutions is to increase the number of Base Stations (BS) with each covering a small area. But, increasing the number of BSs requires high deployment cost. Hence, a cost effective solution is needed to cover the required area while providing desired Signal to Interference plus Noise Ratio (SINR) to the users so as to meet the demand of the future cellular networks. To achieve the high data rate wireless services, Orthogonal Frequency Division Multiple Access (OFDMA) is one of the most promising modulation and multiple access techniques for next generation wireless communication networks. In OFDMA, users are dynamically allocated subcarriers and time-slots so that it is possible to minimize co-channel interference from neighboring cell by using different sub-carriers. Therefore, OFDMA based multi-hop system offers efficient reuse of the scare radio spectrum.

We consider an OFDMA-based cellular system in which users arrive and depart dynamically. Each arriving user demands rate $\bar{r}$. If the required rate can be provided, only then a user is accepted, otherwise it is blocked. Depending on the SINR experienced by an arriving user, the BS computes the subcarriers that are needed to be allocated to the user so as to provide the required rate. If the required sub-channels (i.e., a group of subcarriers) are available, then the user is admitted. Note that the SINR decreases as the distance between the BS and the user increases. Thus, the users at the cell boundary can cause blocking probability to be high. Since the number of admitted users is directly proportional to the revenue of the service provider, it is imperative to design solutions that allow accommodating a large number of users. This motivates us to propose a Fixed Relay based cellular network architecture that is well suited to improve the SINR at the cell boundary, and thus can possibly increase the number of admitted users. Relaying is not only efficient in eliminating coverage holes throughout the coverage region, but more importantly; it can also extend the high data rate coverage range of a single BS. Therefore bandwidth and cost effective high data rate coverage may be possible by augmenting the conventional cellular networks with the relaying capability.

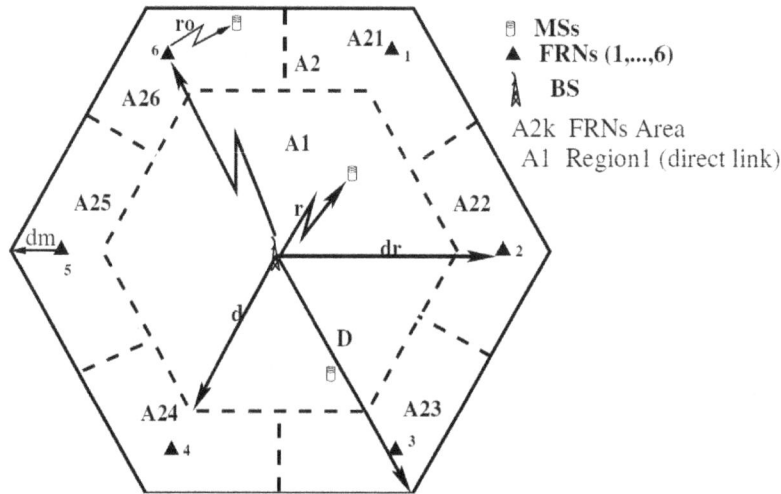

Fig.1. Layout of FRN enhanced Cellular system

We consider a cellular system with six fixed relay nodes (FRNs) that are placed symmetrically around the BS as shown in Fig.1. Mobile stations (MSs) in outer regions A_{21} to A_{26} can use relaying to establish a better path than the direct link to BS. The key design issues in such systems are the following: (1) How sub-channels should be assigned for (a) direct MS to BS links, (b) FRN to BS links, and (c) MS to FRN links. Algorithm for this is referred to as the *channel partitioning scheme*. (2) How sub-channels can be used across various cells. Algorithm for this is referred to as *channel reuse scheme.*

Effective channel partitioning maximizes utilization of every channel in the system, and thus obtains high spectrum efficiency in cellular systems [3]. For a cellular system enhanced with FRNs, the main idea of channel partitioning is to optimally assign the resources to the MS-BS, FRN-BS, and MS-FRN links. Such intra-cell spectrum partitioning along with the channel reuse scheme not only grant the data rate demanded by each user, but also manages the inter-cell interference by controlling the distance between any pair of co-channel links.

The concept of channel partitioning for FRN based cellular system has been discussed in [4], [5], [2], [8], [7], [13]. In [4], full frequency reuse scheme was proposed. The authors divided the cell in to seven parts and allocated six sets of subcarriers to FRN link and the remaining to BS. The authors aimed at exploiting the multiuser diversity gain. In [5], frequency reuse scheme was proposed based on dividing the outer region in to six and also sectoring the inner region. In [2], a pre-configured relay channel selection algorithm is proposed to reuse the channels that are already used in other cells on the links between FRNs and MSs. This scheme may suffer from high co-channel interference on FRN-MS links. In [7], [13], frequency partitioning and reuse schemes for cellular WLAN systems with mobile relay nodes are proposed. Relays are mobile as the MSs themselves act as a relay for other MSs. Since the relay is mobile, the channel between relay and the BS can change, which will result in a large number of inter-relay handoffs. Furthermore, MSs acting as relays may not be cooperative because of the power consumption and the security issues. In [8], a coverage based frequency partitioning scheme is proposed. The scheme assumes that the relay nodes are placed at a distance equal to the two-third of the cell radius, and does not consider optimal relay placement. Some other proposals for frequency management include the use unlicensed spectrum [11], and the use of directional antennas [1].

In our paper [6], we proposed a channel partitioning and a channel reuse scheme for increasing system capacity to support some preceding standards like Global System for Mobile Communications (GSM). In this paper, we extend the channel partitioning and channel reuse scheme for OFDMA cellular networks which results in increased system capacity and spectral efficiency. We also consider the optimal relay placement based on different parameters. We show that with the appropriate relay placement, the system performance can be improved significantly. As a result, the number of users that can be accommodated in the system can be maximized while providing each user with its required rate.

The paper is organized as follows. In Section 2, we describe our system model. In Section 3, we propose our channel partitioning and relay placement scheme based on different parameters. In Section 4, we evaluate the performance of the proposed schemes using numerical computations and simulations. In Section 5, we conclude the paper.

2. SYSTEM MODEL AND DESCRIPTION

2.1. System Configuration

We consider a cellular system consisting of regular hexagonal cells each of edge length D. Each cell has a BS and six FRNs situated symmetrically around the BS at a distance d_r from the center and d_m from the cell edge as shown in Fig.1. Let the total bandwidth available for the uplink is W units. Let the user density in the cell be λ, i.e., in a region of area A, λA MSs are present. Let each MS demands unit rate from the system. We assume the shortest distance routing scheme [9]. For a given MS position, let the distance from BS be d^* and nearest FRN be d^*_m. If $d^* \leq d$, then MS communicates with the BS directly; otherwise it communicates through the nearest relay using two hop route. Because of the specified routing scheme, a cell can be partitioned into seven regions as shown in Fig.1. We define the region covered by BS as the inner region (A_1) and the region covered by FRNs as outer region (A_2). Outer region is further divided into A_{2k}, for k = 1, . . ., 6. All the MSs in region A_1 communicate directly to the BS, and the MSs in k^{th} A_2 region communicate to BS through relay k^{th} FRN.

2.2 Channel Partitioning and Reuse Scheme

The channel partitioning scheme, partitions the uplink bandwidth into thirteen orthogonal segments, viz. W_1, $W_{2,k}$ and $W_{3,k}$ for k = 1, . . . , 6. The band W_1 is used by the MSs in region A_1, the band $W_{2,k}$ is used by k^{th} FRN to communicate with the BS, and the band $W_{3,k}$ is used by the MSs in k^{th} A_2 region to communicate with k^{th} FRN. Let $W_2 = \sum_{k=1}^{6} W_{2,k}$ and $W_3 = \sum_{k=1}^{6} W_{3,k}$. Because of the channel partitioning, there is no intra-cell interference, and the system performance is mainly determined by inter-cell co-channel interference.

For channel reuse scheme, we assume that the frequency reuse distance is 1, i.e., each cell uses the complete bandwidth W for the uplink communication [10]. This may cause significant co-channel interference. But, because of the channel partitioning scheme and the symmetry in the system, we can reduce the co-channel interference by using the following channel reuse scheme.

In each cell, inner region uses the same band W_1. While k^{th} FRN uses band $W_{2,k}$ to communicate with BS and MS in k^{th} A_2 region uses W_{3k} band to communicate with k^{th} FRN.

2.3. Propagation Model

Wireless channel suffers from fading. Fading is mainly divided into two types, slow and fast. Slow fading is due to path-loss and shadowing, while fast fading is due to multi-path. In this paper, we assume that the code lengths are large enough to reveal the ergodic nature of fast fading. Hence, we do not explicitly consider multi-path effect. We focus on the path-loss and shadowing in the analysis. Because of the path-loss, the received signal power is inversely proportional to the distance between the transmitter and the receiver. In general, the path-loss P_L between a transmitter and a receiver is given as,

$$P_L = \frac{P_T}{P_R} G_T G_R = \left(\frac{4\pi f}{c}\right)^2 \left(\frac{d^*}{d_0}\right)^{\gamma} \qquad (1)$$

where P_T is the transmitted power; P_R is the received power, G_T and G_R are the antenna gain of transmitter and receiver respectively; f is the carrier frequency, c is the speed of light; d^* is separation between the transmitter and receiver; d_0 is the reference distance, and $\gamma > 0$ is the path-loss exponent [10].

3. RELAY PLACEMENT SCHEME

Improvement in capacity and increase in coverage area are the main benefits of FRNs. These benefits of FRNs are based on the position of relays in the cell. Deploying FRNs around the edge of the cell help the edge users. However, when they are placed at inappropriate locations, may cause interference to the edge users of the neighboring cell. Therefore, optimal placement of FRNs is a key design issue. Consider the downlink scenario where the BS encodes the message and transmits it in the first time slot to nearby MSs and FRNs. FRNs transmits the message to MSs at the cell boundary in the second time slot. FRNs are either Decode-and-Forward (DF) type, which fully decodes and re-encodes the message, or Amplify-and-Forward (AF) type, which amplifies and forwards the Message to MSs in the second hop. Note that the reverse will be for uplink scenario. In both uplink and downlink scenarios, we consider non-transparent type relays, i.e., MSs in the first hop communicate to BS while MSs in the second hop communicate only to FRNs.

3.1 Problem Formulation

In this section, we formulate the problem and describe our approach. Consider uplink scenario in cellular system as shown in Fig.1. In the cell, an MS can communicate directly with the BS through one-hop, or via FRN based on distance routing protocol [9]. Now, we assume that MSs are uniformly distributed. Hence, the average number of users in any region is proportional to its area. Let N_1 be the average number of MSs in A_1 (direct link) and N_2 be the average number of MSs in A_2 (relay link). Clearly, $N_1 = \lambda A_1$ and $N_2 = \lambda A_2$.

Let R_1, R_2 and R_3 be the data rate achieved on MS-BS (direct link), MS-FRN and FRN-BS links respectively. When MS transmits data to BS via FRN, the rate on MS-FRN link (R_3) should be equal to rate on FRN-BS link (R_2), i.e.,

$$R_2 = R_3,$$
$$W_2 log_2(1 + \Gamma^a_{RM}) = W_3 log_2(1 + \Gamma^a_{BR}) \qquad (2)$$

where Γ^a_{RM} and Γ^a_{BR} denote the worst case SIR for links MS to k^{th} FRN and k^{th} FRN to BS given that the MS is in any of the regions A_{2k}'s for k = 1, . . . , 6, respectively. Moreover, the data rate achieved on direct link (R1) is related to the rate on MS-FRN and FRN-BS (R_2 and R_3) as follows:

$$\frac{R_1}{R_2} = \frac{N_1}{N_2} = \frac{A_1}{A_2} = \frac{A_1}{\sum_{i=1}^{6} A_{F,i}} = \frac{\left(\frac{d}{D}\right)^2}{1-\left(\frac{d}{D}\right)^2} ,$$ (3)

where $R_1 = W_1 log_2(1 + \Gamma_{BM}^a)$. Γ_{BM}^a denotes the worst case SIR on a direct link MS to BS given that the MS is in the region A_1.

The worst case SIR is obtained by considering a scenario in which the transmitter of interest is placed at the longest possible distance from its receiver, while its interferers are placed at the shortest possible distance from the receiver. The worst case SIR of a link provides a lower bound on its actual SIR. Thus, to guarantee the required rate on a link, it suffices to guarantee the rate for the worst SIR on the link. We note that the resource allocation depending on the worst SIR may be conservative. But, it is conducive for robust and scalable implementation as the worst SIR does not change on account of system dynamics caused by the arrival and departure of users and also by the user mobility. The co-channel interference to the BS/FRN of interest is assumed to be from MSs or FRNs that links to first tier or upper tier cells (i^{th} MS or k^{th} FRN, the expression for the worst case SIRs of the three links is as follows.

$$\Gamma_{BM}^a = \frac{1}{d^{\gamma b}} \left[\sum_{i=1}^{N} \frac{1}{d_i^{\gamma b}} \right]^{-1}$$ (4)

$$\Gamma_{BR}^a = \frac{1}{d_r^{\gamma r}} \left[\sum_{i=1}^{N} \frac{1}{d_{ri}^{\gamma r}} \right]^{-1}$$ (5)

$$\Gamma_{RM}^a = \frac{1}{(d_m^{\gamma m})} \left[\sum_{i=1}^{N} \frac{1}{d_{mi}^{\gamma m}} \right]^{-1}$$ (6)

where, d is the distance of furthest MS in A_1 from BS of interest (BS_0). Clearly, the distance between a MS in region A_1 and BS_0 is less than or equal to d while d_i is the shortest possible distance between MS of ith interferer cell and BS_0. Similarly d_r is the distance of kth FRN of BS_0 and d_{ri} is the distance between k^{th} FRN of ith interferer cell and the BS_0; d_m is the furthest possible distance of MS from k^{th} FRN within the coverage of BS_0 and d_{mi} is the distance between MS that is within the coverage of kth FRN of ith interferer cell and k^{th} FRN of BS_0. $\gamma b, \gamma r$ and γm are path-loss exponent of MS-BS, FRN-BS, and MS-FRN links, respectively.

The channel partitioning (W_1, W_2, and W_3) for the proposed scheme is obtained by solving Equ. (2) and (3). Let us define,

$$\Delta = \frac{\left(\frac{d}{D}\right)^2}{1-\left(\frac{d}{D}\right)^2} \frac{log_2\left(1+\Gamma_{RM}^a\right)}{log_2\left(1+\Gamma_{BM}^a\right)} + \frac{log_2\left(1+\Gamma_{RM}^a\right)}{log_2\left(1+\Gamma_{BR}^a\right)} + 1.$$

$$W_1 = \frac{log_2(1 + \Gamma_{RM}^a)}{log_2(1 + \Gamma_{BM}^a)} \frac{W}{\Delta};$$ (7)

$$W_2 = \frac{W}{\Delta};$$ (8)

$$W_3 = \frac{log_2(1 + \Gamma_{RM}^a)}{log_2(1 + \Gamma_{BR}^a)} \frac{W}{\Delta}.$$ (9)

Equ.(7), (8) and (9) show that the channel partitioning in each FRN enhanced cell is dependent on the worst case SIRs of the three links which in turn depends on the FRN location (Equ.(4),

(5) and (6)). With this partitioning scheme, channel assignment to MSs can be conducted at both BS and FRNs. The same technique can be applied to share a channel among FRNs.

The aim of channel partitioning and relay positioning scheme is to maximize the number of users while satisfying the required rate of each user. Hence, maximizing the user density to support large number of users and determining the ratio of frequency band of three links are the key optimization problem. Therefore, the optimal user density of the inner and outer region based on SIR is formulated as,

$$maximize\ \lambda$$

$$s.t\ W_1, W_2\ and\ W_3\ satisfying$$
$$Equ.\ (7), (8) and\ (9) \tag{10}$$

Given that the relay nodes are placed at distance $d_r < D$, the worst case SIRs can be explicitly computed using Equ.(4), (5) and (6) and the geometry of the cellular system. Now, the goal is to find d^* which maximizes λ. On solving the Equ. (10), determine d^*, which maximize the number of users supported in the system.

3.2 SIR Computation for the proposed scheme

We compute the SIR for each relay enhanced Cellular network. We consider the interference from the cells of two tires surrounding the BS_0 as shown in Fig.2. In this computation, the desired MS is assumed to be placed at the furthest distance from its BS/FRN while interferer MSs are located at the nearest possible locations to the BS_0/FRN_1 as shown in Fig.2. SIR for BS-MS (direct link) is given by

$$\Gamma^a_{BM}(d) = \frac{1}{d^{\gamma b}}\left[\sum_{i=1}^{18}\frac{1}{d_i^{\gamma b}}\right]^{-1} = \frac{\left(\frac{d}{D}\right)^{-\gamma b}}{6\left(\left[\frac{\sqrt{3}}{2}\left(2-\frac{d}{D}\right)\right]^{-\gamma b} + \left[\frac{\sqrt{3}}{2}\left(4-\frac{d}{D}\right)\right]^{-\gamma b} + \left[\frac{\sqrt{3}}{2}\left(3-\frac{d}{D}\right)\right]^{-\gamma b}\right)} \tag{11}$$

where d_i is the distance between desired BS (BS_0) and interfering MSs of interfering BSs. Now, we consider the SIR computation of BS-FRN link where BS0 communicates with its FRN_1, the SIR for this link depends on the distance of FRN_1 of interferer cell from the FRN_1 of BS_0 as shown in Fig.4.

Let us define

$$\chi_{i,j} = (iD^2 + jd_rD + d_r^2 +)^{-\frac{\gamma r}{2}} + (iD^2 + jd_rD - d_r^2)^{-\frac{\gamma r}{2}}$$

$$\Gamma^a_{BR}(d_r) = \frac{1}{d_r^{\gamma r}}\left[\sum_{i=1}^{N}\frac{1}{d_{ri}^{\gamma r}}\right]^{-1} = d_r^{-\gamma r}\left\{(3D + d_r)^{\gamma r} + (3D - d_r)^{\gamma r} + 2(3D^2 + d_r^2)^{-\frac{\gamma r}{2}} + \right.$$
$$\left. 2\left[\chi_{3,3} + \chi_{12,6} + \chi_{9,3} + (12D^2 + d_r^2)^{-\frac{\gamma r}{2}}\right]\right\}^{-1} \tag{12}$$

Now, we compute the SIR on FRN-MS link. When the interference of the FRN is of concern, the interference is contributed by MSs located in other cells that are using the same channel to communicate with their corresponding FRNs. Fig.4 shows a case where FRN_1 of BS_0 communicate with its MS so that the signal from MS of other FRN_1 will be the source of the interference. SIR for this particular scheme is,

$$\Gamma^a_{RM}(d_m) = \frac{1}{d_m^{\gamma m}}\left[\sum_{i=1}^{18}\frac{1}{d_{mi}^{\gamma m}}\right]^{-1} = \frac{\left(\frac{d_m}{D}\right)^{-\gamma m}}{6\left(\left[\frac{\sqrt{3}}{2}\left(1+\frac{d}{D}\right)\right]^{-\gamma m} + \left[\frac{\sqrt{3}}{2}\left(3+\frac{d}{D}\right)\right]^{-\gamma m} + \left[\frac{\sqrt{3}}{2}\left(2+\frac{d}{D}\right)\right]^{-\gamma m}\right)} \tag{13}$$

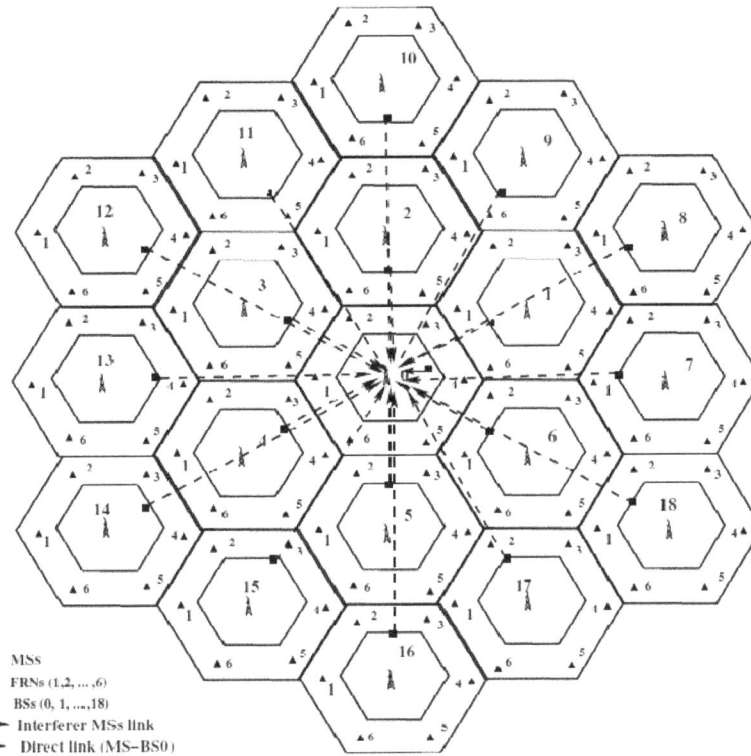

Fig.2. Worst-case interference on MS-BS link (interference received from MS of 18 co-channel cells)

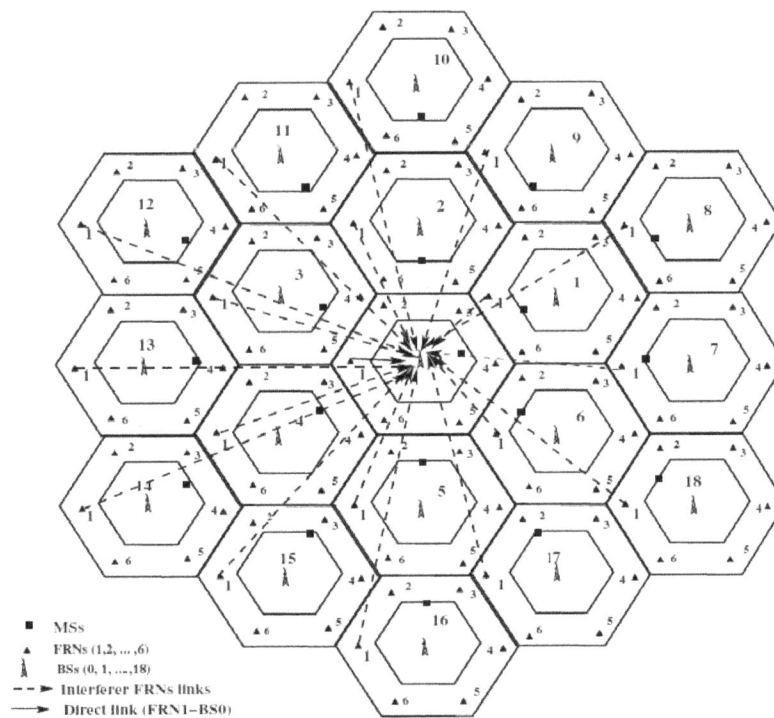

Fig.3. Worst-case interference on BS-FRN link (interference received from FRN1 of 18 co-channel cells)

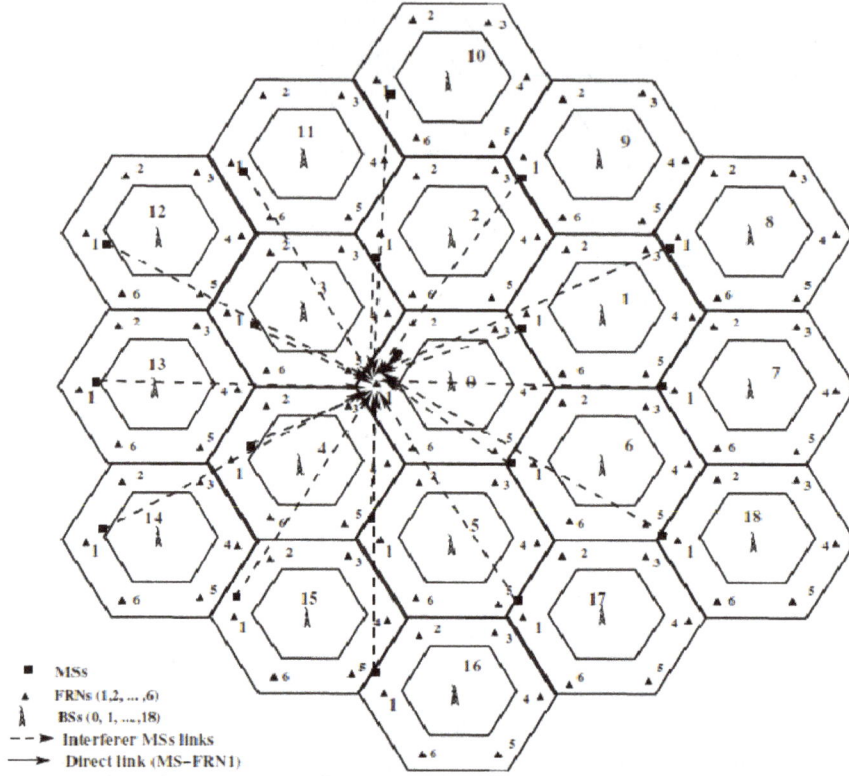

Fig.4 Worst-case interference on MS-FRN link (interference received from MSs in the coverage are of FRN1 of 18 co-channel cells).

3.3. Outage performance analysis

Co-channel interference and shadowing effects are among the major factors that limit the capacity and link quality of a wireless communications system. In this section, we use Gauss-Markov model to evaluate the statistical characteristics of SIR in Multi-hop communication channels. By modeling SIR as log-normally distributed random variable, we investigate the performance of relay placement scheme discussed in Section 3.1. We make a comparison between the above models in terms of performance evaluation where outage probability is a QoS parameter. Further more the result is used to find the more realistic way of channel partitioning and relay FRN placement schemes.

The co-channel uplink interference to the BS/FRN of interest is assumed to be from MSs or FRNs that links to first tier or upper tier cells (MS_i's or FRN_K's). Including the shadowing effect on the three links, the SIR on each link can be described as,

$$\Gamma_{BM}^b = \frac{10^{\xi b/10}}{d^{\gamma b}}\left[\sum_{i=1}^{N}\frac{1}{d_i^{\gamma b}10^{\xi bi/10}}\right]^{-1} = \left[\sum_{i=1}^{N}\left(\frac{d}{d_i}\right)^{\gamma b}10^{\frac{\xi bi-\xi b}{10}}\right]^{-1} \tag{14}$$

$$\Gamma_{BR}^b = \frac{10^{\xi r/10}}{d_r^{\gamma r}}\left[\sum_{i=1}^{N}\frac{1}{d_{ri}^{\gamma r}10^{\xi ri/10}}\right]^{-1} = \left[\sum_{i=1}^{N}\left(\frac{d}{d_{ri}}\right)^{\gamma r}10^{\frac{\xi ri-\xi r}{10}}\right]^{-1} \tag{15}$$

$$\Gamma_{RM}^b = \frac{10^{\xi m/10}}{d_m^{\gamma m}}\left[\sum_{i=1}^{N}\frac{1}{d_{mi}^{\gamma r}10^{\xi mi/10}}\right]^{-1} = \left[\sum_{i=1}^{N}\left(\frac{d}{d_{mi}}\right)^{\gamma m}10^{\frac{\xi mi-\xi m}{10}}\right]^{-1} \tag{16}$$

where d and d_r are the location of desired MS and FRN from the BS_0 on direct link, d_m is the location of desired MS from desired FRN under the second hop. Similarly, d_i and d_{ri} are the location of co-channel interferer MS and FRN from the BS_0, while d_{mi} denotes the location of co-channel interferer MSs from FRN that is associated to BS_0. Shadowing for the desired links are denoted as ξd, ξr and ξm. For interfering links shadowing is expressed as ξdi, ξri and ξmi to denote the interfering link of MS-BS, FRN-BS and MS-FRN; in all cases, $i \in \{1, ..., 18\}$. Let the threshold SIR on BS-MS, BS-FRN and FRN-MS links are denoted as Γ_b, Γ_r and Γ_m respectively. The outage probability on these three links is given as,

$$P_b^{out} = \Pr\left(\Gamma_{BM}^b < \Gamma_b\right) = \int_0^{\Gamma_b} \frac{1}{\sigma_B\sqrt{2\pi}} exp\left[\frac{-(x-m_{\Gamma_B})^2}{2\sigma_B^2}\right] dx = 1 - Q\left(\frac{\Gamma_b - m_{\Gamma_B}}{\sigma_B}\right) \quad (17)$$

where Q(.) is the Gaussian function. We compute the mean m_{Γ_B} and standard deviation σ_B of SIR based on Fenton–Wilkinson's and Schwartz–Yeh's method [12]. Similarly, the outage probability on FRN-BS and MS-FRN are computed as $P_r^{out} = \Pr\left(\Gamma_{BR}^b < \Gamma_r\right)$ and $P_m^{out} = \Pr\left(\Gamma_{RM}^b < \Gamma_b\right)$ respectively. The equivalent SIR experienced by users over two-hop transmission depends on the type of relaying scheme. In this paper we consider Decode-and-Forward Relaying (DF) and Amplify-and-Forward Relaying (AF) relaying schemes.

3.3.1. Decode and Forward Relaying

Consider an uplink scenario, where each FRN decodes the signal received from MS-FRN link and re-transmits to BS on the FRN-BS link. In this scheme, the end to end rate achieved from BS to MS is determined by the minimum rate achieved among the rates of MS-FRN-BS and links, i.e., if the required rate for this scheme is R_{DF}, then

$$2R_{DF} \leq \min\{ log_2\left(1 + \Gamma_{BR}^b\right), log_2\left(1 + \Gamma_{RM}^b\right)\} \quad (18)$$

If the required rate is not achieved on either of the link, then user is said to be in outage. Let the outage probability in DF scheme be P_{DF}^{out}, it can be expressed as

$$P_{DF}^{out} = \Pr\left(\min\{ log_2\left(1 + \Gamma_{BR}^b\right), log_2\left(1 + \Gamma_{RM}^b\right)\} \leq 2R\right) \quad (19)$$

Equivalently, the outage probability on MS-FRN and FRN-BS links can be computed as

$$P_{DF}^{out} = \Pr\left(\Gamma_{BR}^b < \Gamma_r\right)\Pr\left(\Gamma_{RM}^b < \Gamma_m\right) = P_r^{out}.P_m^{out} \quad (20)$$

3.3.2. Amplify and Forward Relaying

In amplify and forward relaying scheme, the AF relay amplifies the analog signal received from the BS and transmits an amplified version of it to the MSs [14]. Let the experienced SIR for this scheme be Γ^{AF}, then it is computed as follows.

$$\Gamma^{AF} = \frac{\Gamma_{BR}^b \Gamma_{RM}^b}{\Gamma_{BR}^b + \Gamma_{RM}^b + 1} \quad (21)$$

Let Γ be the threshold SIR. If the experienced SIR falls below the threshold, the user is said to be in outage and outage probability is given as P_{AF}^{out} out for AF scheme [14].

$$P_{AF}^{out} = \Pr\left(\frac{\Gamma_{BR}^b \Gamma_{RM}^b}{\Gamma_{BR}^b + \Gamma_{RM}^b + 1} \leq \Gamma\right) \quad (22)$$

Note that Shadowing is usually represented by i.i.d. log-normal model in wireless multi-hop models. However, shadowing paths are correlated. In this scheme, we consider correlation among interferers and also the correlation that may exist between interferer and desired signals.

3.3.3. Sectoring the inner region of the cell

Sectoring the inner region of FRN enhanced cell can efficiently reduce the co-channel interference. In sectoring, the total subcarriers in the inner region will be further divided among the number of sectors in the inner region. For example, if 60^0 sectoring is applied, W_1 is partitioned into six orthogonal segments, viz. $W_{1,i}$ for $i = 1, ..., 6$. MSs present in any sector of the inner region can use dedicated subcarriers allocated to that sector. Hence, these orthogonal sub-channels greatly reduce the inter-cell interference. As it was explained in Section 3.1, the total band allocated to inner region depends on the FRN placement scheme.

4. RESULTS AND DISCUSSION

In this section, we present both analytical and simulation results to illustrate the performance of our proposed algorithms. We use Matlab for modeling a cellular network under varying channel conditions. We analyze the performance of channel partitioning and relay placement schemes described in previous sections. The parameters used for simulations are shown in Table 1.

Table 1. List of the simulation parameters

Parameters	Values
Carrier Frequency	5GHz
System Bandwidth (W)	25.6MHz
Cell Radius (D)	1000m
Standard Deviation(σd, σr and σm)	8, 5, 8
Correlation Coefficient	0.5
Path-loss Exponent (γd, γr and γm)	3.5, 2.5, 3.5
BS Transmit Power (P_{BS})	40dBm
FRN Transmit Power (P_{FRN})	20dBm
MS transmit Power (P_{MS})	2dBm
Threshold (Γ)	-10dBm
Thermal Noise (N)	-100dBm

Fig.5 shows the uplink SIR experienced by MS as it moves from the center to the boundary of the cell. The SIR experienced by the MSs of inner region (MS-BS) decreases monotonically while the SIR experienced by the MSs of relaying region (MS-FRN) increases such that the decrement in SIR of the inner region is compensated by SIR experienced by the MSs of outer region. The SIR experienced on the BS-FRN link is fairly constant over the entire range of d/D. This figure, demonstrates that FRN enhanced cellular system offers remarkable improvement in SIR experienced by users over the conventional system without relays. Fig.6 shows the variation of total data rate of the inner (R1) and outer region (R2) as a function of d/D for the path-loss exponent of 3. From this figure, we observe that FRN positioning has significant impact on the performance of FRN based OFDMA cellular network. In addition, the data rate is also dependent on environmental conditions expressed by path-loss exponent.

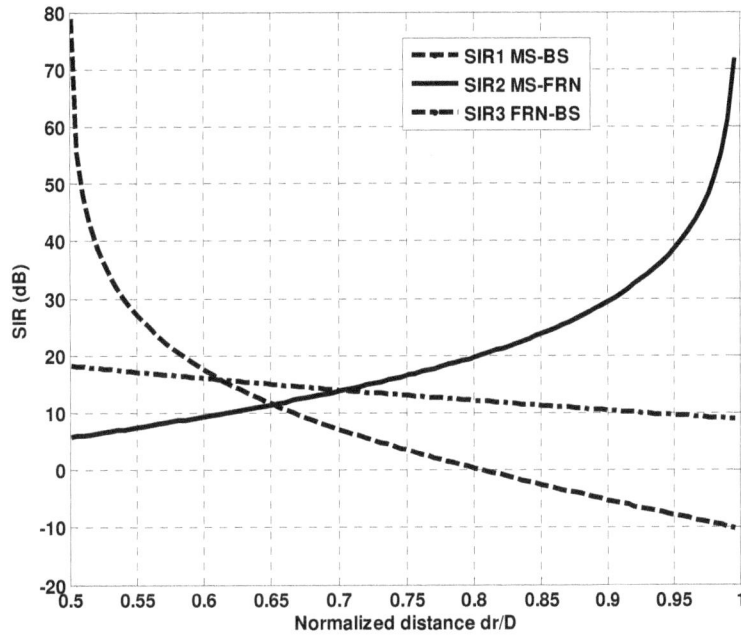

Fig.5 Distribution of SIRs versus dr/D

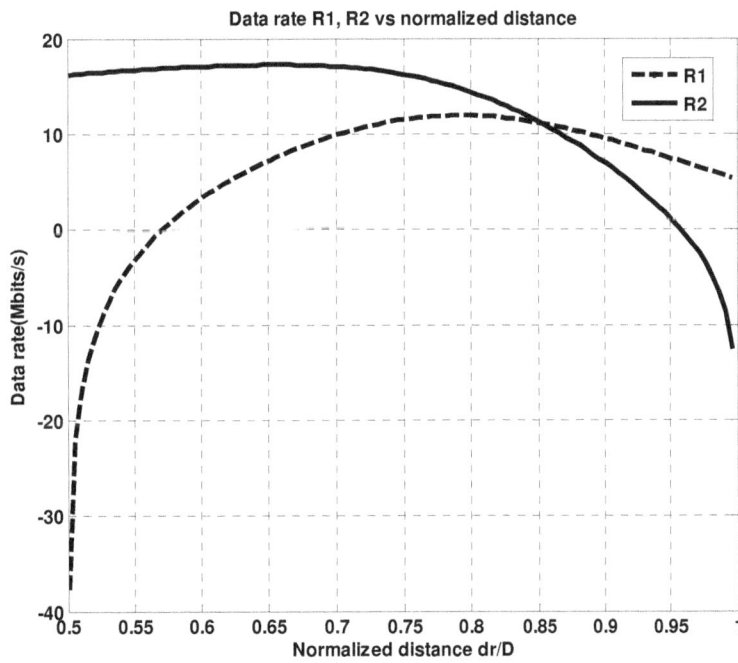

Fig.6 Cell capacity versus dr/D

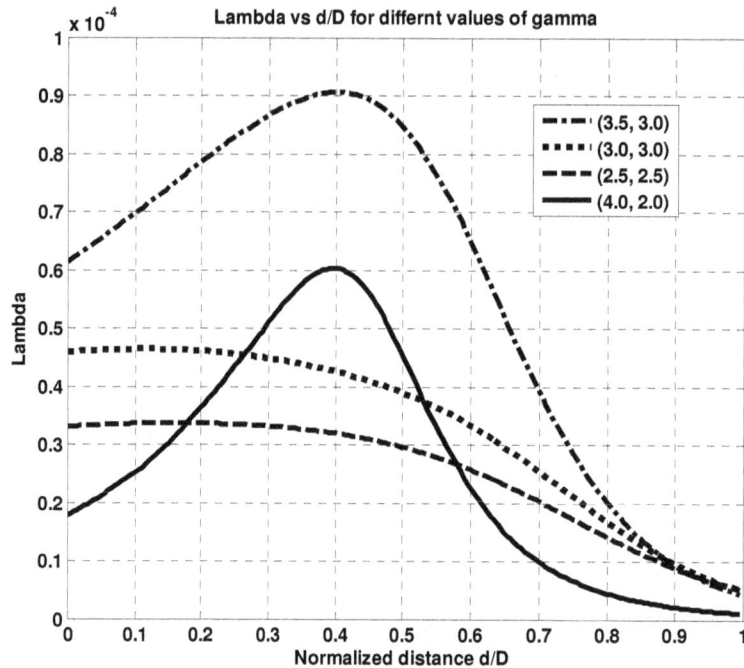

Fig.7 User density for pairs of path-loss exponent (for MS-BS/FRN and FRN-BS link)

Fig.7 shows locations of maximum user density for a given path-loss exponent that indicates the optimal FRN position under different path-loss conditions. The plots are denoted by pair of path-loss exponent of two links. The first number indicates the path-loss exponent of MS-BS/FRN link and second indicates that of FRN-BS link. From this figure we observe that for path-loss exponent of (3.5, 3) and (4, 2), the optimal relay node location is at 70% of the cell radius. In case of (3, 3) and (2.5, 2.5), the optimal FRN position is 55% of the cell radius. It can be seen from this figure that the optimal FRN location depend on path-loss exponents which contradict with the assumption of FRN location as dr/D = 2/3, i.e., d/D = 1/3 in [8] and other literatures.

Fig.8 shows the partitioning of available uplink frequency band (W_1, W2 and W_3) in each FRN enhanced cell to serve three set of links (MS-BS, FRN-BS and MS-FRN links). From this graph, we observe that the amount of band shared by these three set of links is dependent on FRN location. Such spectrum partitioning scheme can be conducted at both BS and FRNs. Fig.9 shows the effect of path-loss exponent on channel band partitioning for a fixed distance of MS and FRN/BS. We consider path-loss exponents of 2.0 and 2.5 for FRN-BS link and determine the partition (W_1, W_2, and W_3). For a fixed radial distance of a user from BS, the SIR experienced by user decreases as path-loss exponent increases. This intern affect the frequency band allocation to each link as shown in Fig.9.

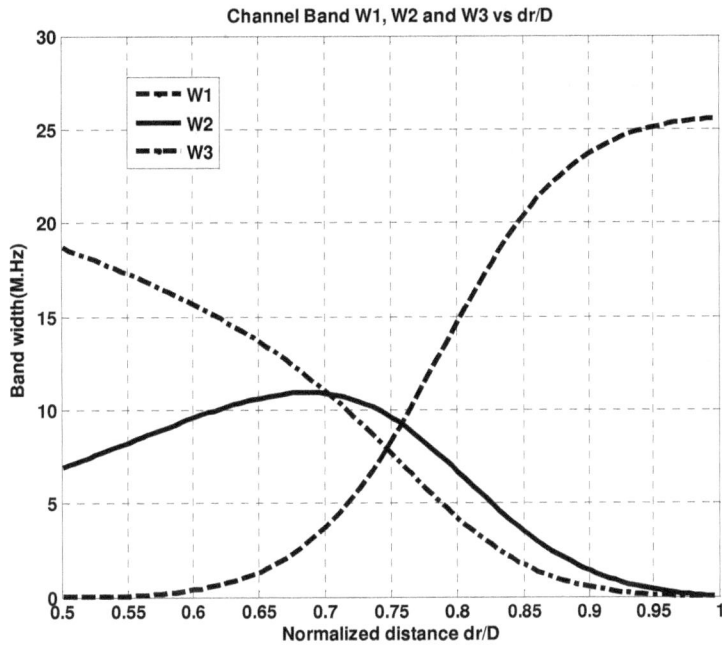

Fig.8 Effect of dr/D on channel Band

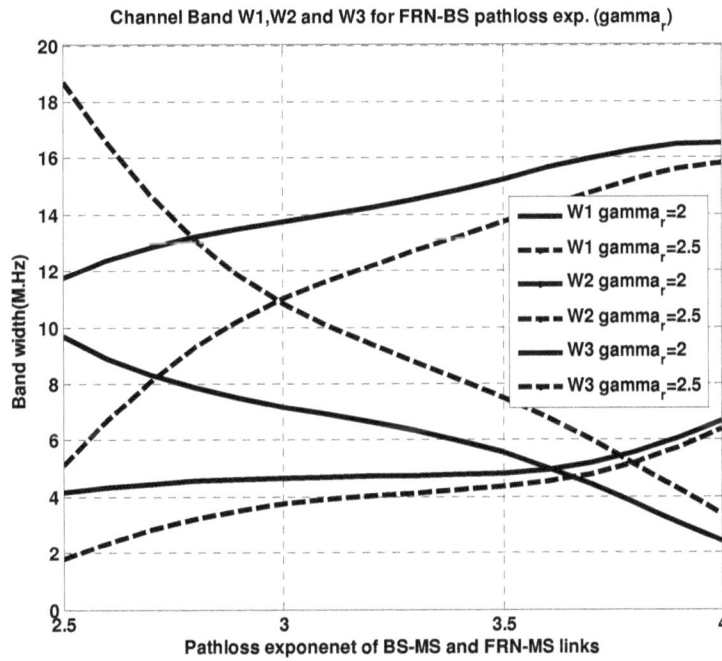

Fig. 9 Effect of path-loss exponent on channel Band

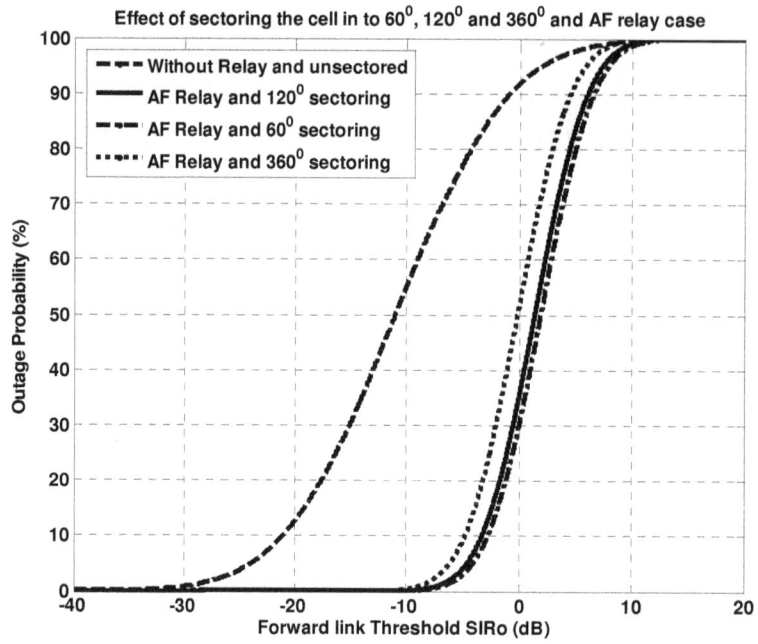

Fig.10. Outage probability AF Relay compared with cellular networks without relay; the inner region of relay enhance cell with 60^0, 120^0 and 0^0 sectoring.

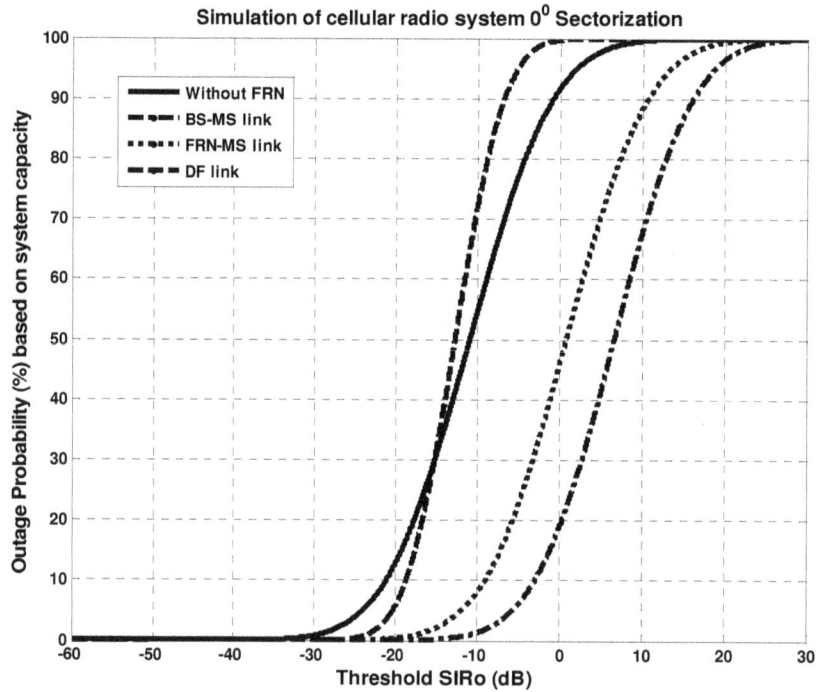

Fig.11. Outage probability of DF Relay compared with cellular networks without relay; no sector in inner region.

Simulation of cellular radio system 120^0 Sectorization

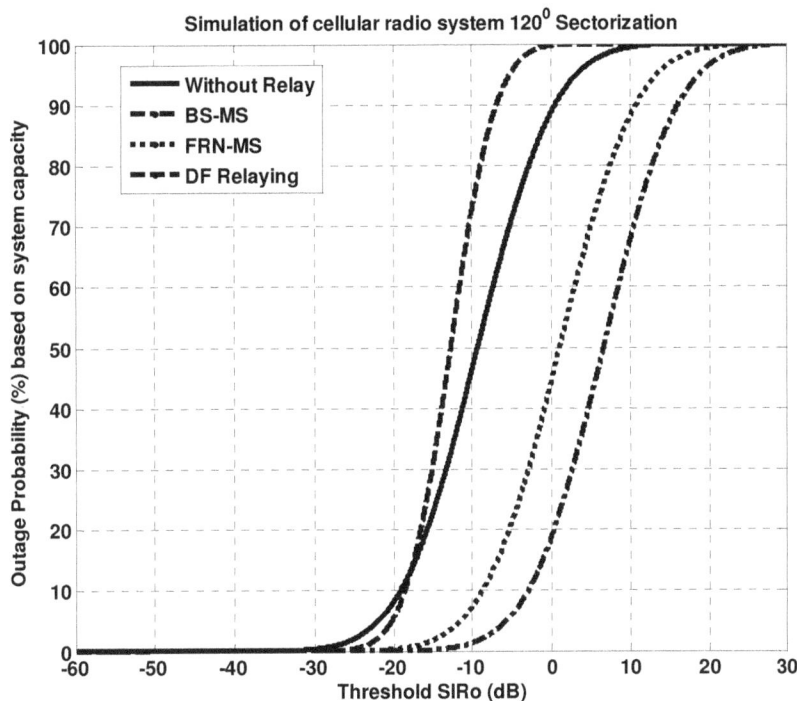

Fig.12. Outage probability of DF Relay compared with cellular networks without relay; 120^0 sector in the inner region.

Fig.10 shows the outage probability versus SIR in dB for an AF relay and without relay. From the figure, we observe that AF relay scheme achieves significant improvement on the performance of the cellular network. It reduces the outage probability significantly. For example, at the threshold SIR of 0 dB, the AF scheme can improve the outage probability from 90% to 30%. In addition, sectoring of the inner cell also contribute in the overall system improvement

Fig. 11 shows the outage probability for DF relays. For lower value of threshold SIR, both AR and DF relays show significant improvement on the outage performance of the cell. However, for the higher values of SIR threshold limit, the outage probability increases. This shows that the system is dominated by interference. At lower region of SIR threshold, AF relaying scheme performs better than DF in improving the outage probability of the users in terms of link quality. Fig.11 and Fig.12 compares the outage probability of a cell using DF relaying scheme by sectoring the inner region of the cell. Even though sectoring improves the outage probability of users, there is a trade-off that it may degrade the capacity of a system in terms of maximum carried traffic.

5. CONCLUSION

In this work we propose new techniques for channel partitioning and FRN positioning schemes in cellular OFDMA network. We investigate channel partitioning and FRN placement under different path loss conditions. Our analytical results show that the optimal relay node placement depends on the path-loss exponent of the environment. For example it is at 0.7 of dr/D ratio for path-loss exponent of (3, 3). In addition to this, simulation results show that our proposed schemes improve the system capacity and provides better QoS for users present at the boundary of cells.

ACKNOWLEDGMENT

This research work is supported by India-UK Advanced Technology Centre (IU-ATC) of Excellence in Next Generation Networks Systems and Services.

References

[1]   Zaher Dawy, Sami Arayssi, Ibrahim Abdel Nabi, and Ahmad Husseini. Fixed relaying with advanced antennas for CDMA Cellular Networks. In IEEE GLOBECOM proceedings, 2006.

[2]   H. Hu, H. Yanikomeroglu, and et al. Range extension without capacity penalty in Cellular Networks with digital fixed relays. IEEE Globecom, Dec 2004.

[3]   I. Katzela and M. Naghshineh. Channel assignment scheme for cellular mobile telecommunication systems: A comprehensive survey. IEEE Personal Communication, pages 10–31, June 1996.

[4]   Jian Liang, Hui Yin, Haokai Chen, Zhongnian Li, and Shouyin Liu. A novel dynamic full frequency reuse scheme in OFDMA cellular relay networks. In Vehicular Technology Conference (VTC Fall), 2011 IEEE, pages 1 –5, Sept. 2011.

[5]   Min Liang, Fang Liu, Zhe Chen, Ya Feng Wang, and Da Cheng Yang. A novel frequency reuse scheme for OFDMA based Relay enhanced cellular networks. In Vehicular Technology Conference, 2009. VTC Spring 2009. IEEE 69th, pages 1 –5, April 2009.

[6]   Sultan F. Meko and Prasanna Chaporkar. Channel Partitioning and Relay Placement in Multi-hop Cellular Networks. In Proc. IEEE ISWCS, 7-10 Sept. 2009.

[7]   S. Mengesha, H. Karl, and A. Wolisz. Capacity increase of multi-hop cellular WLANs exploiting data rate adaptation and frequency recycling. In MedHocNet, June 2004.

[8]   M. Rong et al P. Li. Reuse partitioning based frequency planning for relay enhanced cellular system with NLOS BS-Relay links. IEEE, 2006.

[9]   V. Sreng, H. Yanikomeroglu, and D. D. Falconer. Relayer selection strategies in cellular networks with peer-to-peer relaying. IEEE, VTC, 03, Oct 2003.

[10]   D. Tse and P. Viswanath. Fundamentals of Wireless Communications. Cambridge University Press, 2005.

[11]   D. Walsh. Two-hop relaying in CDMA networks using unlicensened bands. Master's thesis, Carleton Univ., Jan 2004.

[12]   Jingxian Wu, N.B. Mehta, and Jin Zhang. Flexible lognormal sum approximation method. In EEE GLOBECOM, pages 3413–3417, Dec 2005.

[13]   H. Yanikomeroglu. Fixed and mobile relaying technologies for cellular networks. In Second Workshop on Applications and Services in Wireless Networks, July 2002.

[14]   Bhuvan Modi1, A. Annamalai1, O. Olabiyi1 and R. Chembil Palat. Ergodic capacity analysis of cooperative Amplify-and-Forward relay networks over rice and nakagami fading channels. International Journal of Wireless & Mobile Networks (IJWMN) Vol. 4, No. 1, February 2012.

9

FED: Fuzzy Event Detection model for Wireless Sensor Networks

HadiTabatabaee Malazi[1],Kamran Zamanifar[1] and Stefan Dulman[2]

[1]Department of Computer Eng., University of Isfahan, Iran
{tabatabaee, zamanifar}@eng.ui.ac.ir
[2]Embedded Software Group, Delft University of Technology, The Netherlands
s.o.dulman@tudelft.nl

ABSTRACT

Event detection is one of the required services in sensor network applications such as environmental monitoring and object tracking. Composite event detection faces several challenges.The first challenge is uncertainty caused by variety of factors, while the second one is heterogeneity of sensor nodes in sensing capabilities. Finally, distributed detection,which is vital to facilitate uncovering composite events in large scale sensor networks, is challenging.We devised a new fuzzy event detection model which is called FED that benefits from fuzzy variables to measure the intensity as well as the occurrenceof detected events. FED uses fuzzy rules to define composite eventsto enhance handling uncertainty. Moreover, FED provides a node level knowledge abstraction, which offers flexibility in applying heterogeneous sensors. The model is also applicable to a clustered network for distributed event detection. The simulation results show that FED is less sensitive to environmental noise and performs better in terms of percentage of detected eventscompare to a similar approach.

KEYWORDS

Wireless sensor networks, Composite event detection, Fuzzy event detection (FED), Heterogeneity, Uncertainty

1. INTRODUCTION

Event detection is a popular service in environmental monitoring [1]–[3] and object tracking [4] applications.Ambulatory medical monitoring [5], vehicle tracking [6], [7] and military surveillance [8] are some sample applicationsthat event detection plays a key role. The popularity of this service is not limited to the applicationlayer. Several wireless senor network middlewares [9]–[14] provide the required primitives, such as event notification to facilitateevent detection tasks in various applications. "Event detection is a way to dig meaningful information out of thehuge volume of data produced" mentioned S.Li in [15].

Events are generally categorized into *simple* (*atomic*) and *composite* (*complex*) ones. Simple events can be detectedby an individual sensor type, whenever the sensed value is above/below the predefined threshold, while compositeevents (CE) are those that cannot be detected by a single sensor type and require collaboration among various types.

Composite event detection poses several challenges. One of the issues is the effect of uncertainty in the detectionprocess. Environmental noise, message collision, and hardware malfunctioning are some of the factors that maycause uncertainty. Uncertainty sources not only affect the detection of simple events at node level, but they alsoaffect CE detection in fusion points causing both false negatives and false positives. The node density also introducessome challenges. A low node density increases the chance that none of the notifications reach the fusion point,while a higher density introduces a collision problem when nodes attempt to transmit simultaneously.

Event detection applications may use variety of different sensor types to uncover composite events using heterogeneoussensor nodes. The main reasons for applying heterogeneous sensor nodes are hardware constraints and energyconsiderations. Therefore, each node may not be able to detect a composite event based on its local observations.Consequently heterogeneous nodes are required to collaboratively detect composite evens. For example they maysubmit the detected simple events to an aggregation point to detect composite events [16].

The growing trend toward cyber-physical systems [17]–[19] introduces new desired properties such as *knowledgeabstraction* and *in network processing*. Event notification part of the traditional event detection systems does notcompletely fulfil these properties. The basic form of an event notification is a tuple consist of *event name*, *reportingnode ID* and *detection timestamp* which only reports the occurrence of an event in the binary form of *true / false*.To provide more information on the detected event, the sensed value field can be added to the notification tuple. Theproblem with the latter form of notification format is that the fusion point should have the knowledge to interpretthe sensed value to estimate the intensity of the reported event which leads to spread the interpretation knowledgeof sensing values. Consequently, it results in reducing flexibility especially in heterogeneous sensor networks sincemodifying the threshold of the sensed values should be applied in many nodes. Moreover, it puts the burden ofinterpreting the sensed value of the event fusion point which leads to decrease of inside network processing.

Energy efficiency is also one of the main challenges for most of the sensor network services and applications.Traditionally, nodes submit the sensed values of interest to the base station for information fusion. The centralapproach is prone to several shortcomings such as overspending bandwidth and higher energy consumption, sincenearby nodes transmit the same event redundantly. Reducing the number of message transfers has a considerableimpact on the energy consumption of sensor nodes. The alternative approach is to use distributed event detectionby contributing several fusion points such as cluster heads in a clustered network. Distributed event detection alsofaces several challenges such as dynamic topology and diversity of available sensor types in each cluster.

The last challenge is the scalability and dynamic nature of large wireless sensor networks. Considering a clusterednetwork, various types of sensor nodes may join or leave a cluster making it difficult for a cluster head to keep trackof available sensor types especially in networks with large clusters. A mechanism is required to provide the densityof available sensor types in the cluster. The information will help adaptive event detection, since each cluster head(event fusion point) will make a more accurate decision to either wait for another event type, or report a compositeevent based on previously received events.

There are considerable amount of published papers that tackle the challenges from different perspectives. Wecategorize them into four groups. Application specific event detection approaches [5], [7], [8], [20] are the firstcategory that target issues like energy efficiency, accuracy, and application specific challenges. Their main goal is todevised application-specific tailored solutions. Second category of researches attempts to provide required primitivessuch as, efficient notification service mechanism in the middleware layer [11], [12], [14] to enhance event detectionapplications. They usually consider idealistic models for example in communication and do not address the possibleminor problems such as false positive detection of congestion of communication links. The third group of researchesaddress the uncertainty [11], [15], [21] in event detection, and finally the last category of researches focus ondistributed approach of event detection [22]–[24]. The main goal is to reach a consensus between detecting nodesin an efficient way in terms of energy and accuracy.

In this paper we devised a generalized model called *FED* for composite event detection. FED benefitsfrom fuzzy modelling in several ways.*FED* applies fuzzy variables to report simple events and their intensity in anabstracted format. Fuzzy membership functions are used for each sensor type to map the sensed values to fuzzyones. Therefore, the fusion points do not need the interpretation knowledge of individual sensor types resulting in asimplification of using heterogeneous sensors. Fuzzy operators are applied to aggregate the reported events. We alsodefine composite events as fuzzy rules. *FED* is fully compatible with our previously designed clustering scheme,called *DEC* [25]. It can also be integrated with our density estimation algorithm [26] to support clusters with theinformation on available sensor types. We evaluated the approach in different node densities, environmental noise,and sensor false detection rate. The results support the idea that *FED* is less sensitive to uncertainty sources. Thedevised fuzzy model can be applied in distributed detection for a clustered network where event notifications areaggregated in cluster heads.

The rest of the paper is organized as follows. In the next section we present the related work. The distributedcomposite event detection problem is formally discussed in Section 3. *FED* model is introduced in detail in Section4. In Section 5 the model is evaluated and finally we conclude in Section 6.

2. RELATED WORK

A wide range of research has been published on event detection in wireless sensor networks. The focus of attention varies from application specific detection to enhancement of middlewares. Some concentrate on uncertainty in event detection while other devise approaches for distributed aspect of detection. In the following we briefly review some of them.

2.1. Application specific event detection

Some of the published research is dedicated to detect events in a particular application such as vehicle tracking,medical diagnosis and military surveillance.

Keally et al. in [7] devised an event detection framework to fulfil user specific requirements mostly on objecttracking. The framework explores the sensing capability of nodes firstly, to perform collaboration between nodes tomeet the required accuracy based on user demands efficiently and secondly, to change detection capabilities basedon runtime observation adaptively.

Hill et al. in [20] reported their experiment in predicting possible events, based on monitoring and analysing a received stream of data sensed by thousands of sensors in an oil field. They introduced an infrastructure foranalysing event detection by real time monitoring in order to detect possible failures. The framework uses fourtiers including, *user tier*, *early event warning tier*, *sensor publisher tier* and *ontology tier* to address the challengessuch as fast response time, maintaining a long history of events, and combining reported events.

Shih et al. in [5] devised an automated approach for detecting seizure in epilepsy patients. Apart from medicalrequirements, building a light weight device with fewer electrodes is considered as requirements for the target system.The use of wireless technology helps in omitting wires which results in lighter devices. They apply machine learningtechniques such as a *support vector machine* (SVM) classifier to construct reduced channel detectors. Consequentlythe seizure is detected with fewer electrodes.

Tian He et al. in [8] design a military surveillance system that enhances a group of sensor devices to detectand track the positions of moving vehicles cooperatively. The main goal is to alert the command and control unitwhenever an event of interest such as moving vehicle happens. Four major requirements are considered for thetarget approach including, longevity, adjustable sensitivity, stealthiness and effectiveness in precision and locationestimate.

The aforementioned research concentrates on a specific application and devises the solutions based on specificapplication conditions. Consequently they do not provide a generic solution which can be applicable to most of theapplications.

2.2. Middlewares for event detection

Some of the researches concentrate on devising middleware [9], [10], [12], [14] to facilitate applications forefficiently reporting the detected events. In the following, we briefly review features devised by middlewares, suchas *TinyDB* [9], *Impala* [10], and *Mires* [12].

Event based query is one of the facilities that *TinyDB* [9] provides for event detection applications. This type ofquery is triggered whenever an explicit event has happened. In other words, based on the sensed value the specifiedevent will raise an interrupt and the query will be executed. In order to use this facility the programmer should writea component to introduce the event and the signals. The defined events can be further used in queries wheneverrequired.

Impala [10] provides an event based programming model for applications. It assigns a specific middleware agentcalled event filter to fulfil the programming model requirements. The event filter agent is responsible for capturingand dispatching detected events to other middleware agents as well as applications.

Souto et al. in [12] devised a *publish/subscribe* middleware called *Mires*. It provides primitives for publishingdetected events for the subscribed nodes. The publish/subscribe approach used in *Mires* provides an asynchronouscommunication between the elements of a network. This is a valuable advantage in a dynamic nature of WSN. Theevent detection mechanism in *Mires* has three phases. In the first phase nodes advertise their sensing capabilitiesas *Topics*. The advertised messages are then sent to the sink node via a multihop routing protocol. In the secondphase user applications connect to the sink and subscribe those sensing capabilities which they are interested in.In the last phase subscribed messages broadcast down the network. Receiving the subscribed messages, nodes cannotify the detected events (*topics*).

Middleware usually addresses the node level event capturing and dispatching. They provide a programming modelto raise events, which are usually simple events, based on the sensed values. The distribution and aggregation ofthese events is the second aspect of these middlewares. On the other hand they do not address the detection ofcomposite events. Besides, they usually do not specify the architecture for distributed detection. Consequently, theseaspects are mainly forwarded to application layer.

2.3. Uncertainty in event detection

Several approaches have been devised so far to cope with the effect of uncertainty in event detection.

Heinzelman et al. [11] has introduced a proactive service oriented WSN middleware called *MiLAN*. One of theinteresting aspects of the middlware is the capability of switching between sensors with different sensing accuracy.*MiLAN* is able to handle heterogeneous nodes with different sensing accuracy (*Quality of Service*). Applicationsrunning above the middleware layer are powered by the capability to identify their accuracy needs based onapplication states. Generally, uncertainty in event detection contains wider range of issues than *QoS*. *MiLAN* isa remarkable research in dynamically handling accuracy in sensing but it does not address issues such as falsepositive detection, event notification loss and aggregating uncertainties.

S.Li et al. in [15] has defined an event detection service (*DSWare*) using a data centric approach. It supportsdetecting CEs in a sensor network with heterogeneous nodes. An application program can register events bysubmitting an *SQL-like* statement to a group of specified nodes. In order to address CEs, sub-event sets are definedin the statement. The definition of a sub-event consists of several parameters, such as a confidence function anda minimum confidence value for detecting it. To cope with uncertainty *DSWare* uses confidence functions. Aconfidence function takes occurrence of sub-events, in a Boolean data type format, as an input parameter andcomputes a numeric value showing how likely the CE has happened. *DSWare* aggregates the reported events alongthe path to the sink. Consequently, it is not applicable for a clustered network. It also does not provide node levelabstraction in interpreting the sensed values.

Ambiguity in knowledge acquisition for defining composite events is the issue that Manjunatha addresses in[27]. According to the proposed approach, sensors submit their sensed values to an aggregation point. The meanof transmitted values are considered as aggregated value. Then the aggregated value is fuzzified and the inferenceengine looks for any possible composite event, defined as fuzzy rules. Although the approach has some similaritieswith our work, our model differs on several points from [27] in several points. Firstly, Manjunatha in [27] does notaddress false positive detection issues. Secondly, the proposed approach does not consider unreliable communicationand message loss which may cause uncertainty in event detection. Thirdly, sensor nodes transmit the sensed values,which violate node level abstraction in interpreting. Besides, the aggregation method is not appropriate for falsedetection situations. Finally, it seems that the inference engine does not consider time and location correlation indetecting composite events.

Samarasooriya et al. [21] have introduced a fuzzy modelling approach in dealing with uncertainty. The mainfocus is the varying degrees of accuracy in local sensors, specifically the local sensors error probabilities whichare varying in time in a non-random fashion. In other words, they target node level uncertainty in detecting events.They modelled the error probabilities with fuzzy quantities using membership functions. They used a probabilisticapproach to fuse the local sensor decisions and formally prove the performance of their model. Although theytry to model node level error probability, they do not devise a solution at fusion point which includes unreliablecommunication.

2.4. Distributed event detection

From distributed detection point of view,several remarkable researches published so far.

Viswanathan et al. in [22] analyze several distributed detection (*distributed signal processing*) architectures byapplying *Neyman-Pearson formulation*. They investigate the computational complexity in achieving the optimalsolution. Parallel topology with/without fusion center as well as serial and tree topology were studied. They comparethe serial architecture, in which each node makes a decision based on its observation as well as the received decisionsfrom its neighbors and then forwards its decision to the next node in a serial way. One of the important outcomesof the research is, for the case where large number of participate in the distributed detection process, the probabilityof missing event goes to zero with a much slower pace in the serial architecture compared to the parallel one.

Kumar et al. in [28] devised a framework for developing distributed data fusion applications called *DFuse*. Itconcentrates on two main aspects. The first one is providing a wide range of *fusion APIs* to facilitate applicationsin complex information aggregation such as video streams. The second characteristic is the distributed algorithmfor placement of fusion function. The main goal is to find out the optimum placement of fusion point to minimizecommunication cost. *DFuse* provides a heuristic approach to choose a suitable fusion point based on predefinedcost function. The placement process re-evaluated periodically to address network dynamics.

3. PROBLEM DEFINITION

We consider a network consist of m heterogeneous nodes, each node may have different subset of availablesensors.

$$N = \{n_i : i \in m\} \tag{1}$$

Let v be the available sensor types in the network and C_i the capability tuple of node i. Each element in the $C_i$represents a flag for a sensor type. The value of one for s_k indicates that the node is equipped with sensor type kand value of zero shows nonexistence.

$$C_i = (s_k : k \in v) \tag{2}$$

Each node sends its observation upon detection of a simple event. Let y_i be a reported local observation ofsensor node i and u be the global (fusion point) decision on composite event detection. The Eq. 3 shows themapping of local decisions to the composite event detection.

$$u = \gamma_0\big(y_1(.), y_2(.), \dots, y_m(.)\big) \tag{3}$$

Let $\Gamma(\gamma_0, \gamma_1, \gamma_2, \dots)$ be the set of rules that defines composite events, where $\gamma_0(.)$ is the definition of a samplecomposite event. Considering the following probabilities, the goal of the model is to increase $\Lambda(u)$ which is thelikelihood of detection. In the realistic model transmitted observations may fail to reach the destination due tomessage loss. Besides, sensor nodes may have false positive detection due to various reasons including hardwaremalfunctioning.

P_F: $P(u = 1 | H_0)$: global false alarm probability

P_D: $P(u = 1 | H_1)$: global detection probability

P_M: $P(u = 0 | H_1)$: global miss probability

$$\Lambda(u) = \frac{P(u \mid H_1)}{P(u \mid H_0)} \tag{4}$$

In general three conditions are required to detect any CE. These conditions are:

1) In order to detect any CE, a group of predefined simple event types must have been detected. That is, a CE is defined as a set of simple event types.
2) The occurrence timestamp of the reported simple events should be within a limited time period (called *detection window*), which has been defined in the CE definition.
3) The nodes, reporting the simple events should be close enough in terms of geographical distance. That is, in order to infer any correlation among simple events, there should be rational physical distance among reporting nodes. The rational distance between nodes can be calculated based on the sensing range of each sensor. It can also be measured in terms of hop counts.

Given the above detection requirements, the devised approach should fulfil the following goals.

- Detect composite events in the presence of message loss and false positive simple event detection.
- Support heterogeneity of sensor nodes in terms of sensing capability.

- Be expendable for distribute detection in large scale networks.

4. FED MODEL

In this section we describe different aspects of the model. First, we describe the network architecture and in thesecond part, we explain event notification. In the third part, we present composite event detection and finally, weaddress uncertainty problem.

4.1. Network architecture

FED uses clustered network architecture to perform distributed event detection and prevent redundant submissionof simple events to sink node. The cluster heads have the responsibility of aggregating the reported events withintheir clusters and forward the detected composite events to the sink node.

Choosing an appropriate clustering scheme is an important issue in the efficiency of the model. Recall from thedefinition of composite events, a variety of sensor types are required to collaborate in order to detect a compositeevent. Therefore, the clustering scheme should maximize the diversity of sensor types in each cluster to increasethe cluster capabilities in detecting composite events. Traditional clustering schemes such as [29]–[34] fail to fullysatisfy this requirement. They usually do not consider sensor diversity as a clustering parameter.

We have introduced a diversity base clustering scheme called *DEC* [25] to increase the capability of detectingvarious composite events. *DEC* performs four phases, which are *initialization*, *clusterjoining*, *migration*, and*termination*, to generate clusters with maximum possible diversity of sensor types. It applies a cost functionthat uses the residual energy level of a node and the diversity of its neighboring nodes, to elect a cluster head. Tohandle the dynamic nature of wireless sensor networks, *migration* phase of the algorithm has the duty of balancingthe capabilities of the cluster in case where some sensors fail. That is, whenever a scare sensor type fails, thecluster head invites nodes with the same sensor type to migrate. The invitation will be accepted, if it is granted bythe node's current cluster head. The simulation results in [25] show that it produces clusters with higher sensingcapabilities for enhancing event detection applications. For more detailed information on *DEC* please refer to [25].

It should be added that there is also an on-going research to devise a clustering scheme for the mobile nodes toadaptively maximize sensor diversity in each cluster. The first step is to estimate the diversity of sensor types [26]in a mobile network. The estimation will be further used in order to provide the required clustering scheme for themobile network.

4.2. Simple events

Simple event notification is the building block of *FED*. We apply fuzzy logic [35] to provide a node levelabstraction on interpreting the meaning of the sensed values. Based on *FED* model, we fuzzify the sensed valuesinside the sensing node. One of the advantages is to provide a node level abstraction on the meaning of the value.Therefore, the aggregation point is not required to have a full knowledge over all sensor types in the heterogeneoussensor networks.

In *FED* each sensor type is associated with a fuzzy variable and consequently each fuzzy variable has severalfuzzy values. The values are defined based on the specified simple events by the application. Although submittingthe *fuzzy value* of the detected simple event is adequate for applications that only need simple event detection,composite event detection applications need more information on the membership degree of the fuzzy value inorder to aggregate the simple event notifications.

To convert the sensed values into fuzzy ones, a membership function is required. Figure 1 shows a samplemembership function for a heat sensor. The X axis is the temperature degree in Celsius while Y axis shows themembership degree. Values below 26 are out of concern as the threshold is set to be 26°C.

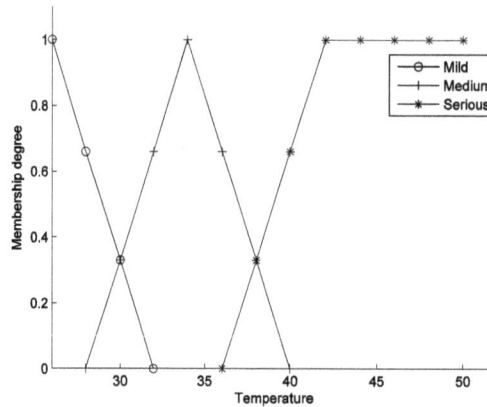

Figure1. Sample fuzzy membership function for heat sensor

The next step is to prepare the event notification message. Simple event notifications in *FED* consist of fivefields. Some of these fields are similar to the traditional event notification format. Equation 5 shows the simpleevent notification in *FED* where e_{name}, n_{id}, t, f_{value} and $d_{membership}$ are event name, node ID, event time, fuzzyvalue and membership degree respectively.

$$y_i = (e_{name}, n_{id}, t, f_{value}, d_{membership}) \tag{5}$$

For example, in an indoor fire detection application the temperature above 26° should be reported. Consider anode with the ID of 43 that has sensed the temperature of 37°at 12:23:39. The following event notification willbe reported. *Temp* shows that a temperature event has happened. The second field shows the ID of the reportingnode while the third element shows the time (more precise time formats can be used to present timestamp field) inwhich the event has been detected. The fuzzy value and related membership degree are the fourth and fifth fieldsrespectively.

$$event = (Temp, 43, 12:23:39 \ GMT, serious, 0.20) \tag{6}$$

4.3. Composite events

Recall from Section 3, the occurrence of a set of predefined simple events is one of the conditions of detectingcomposite events. Considering the fact that simple events are reported as fuzzy notifications, we define compositeevents as fuzzy rules. Here is a sample composite event that has been defined in a fuzzy rule format.

γ_j:*If Heat is MEDIUM and Humidity is LOW and Smoke is Medium then Fire is SERIOUS*

The IF clause shows the set of simple events required to detect the composite event, and the THEN clausespecifies the composite event name. In the composite event definition, each simple event is associated with a fuzzyvalue. The values are used to estimate the likelihood of the composite event.

FED uses two aggregation methods to fuse the transmitted simple events. The first method is used to fuse thesimple events of the same type and the second one is used for the final aggregation. The fuzzy operator OR is usedto aggregate same type simple events.

The next step is to investigate if the possibly correlated notifications can satisfy any fuzzy rule. In order tocalculate the occurrence degree of the detected CE, notifications are aggregated weighted average function. Theactual weight for each simple event type of a specific fuzzy rule (composite event) is chosen based on application requirements. The more important event types will weight higher compared to less important ones. It is also possibleto adaptively choose the weights based on the redundancy of each simple event that has been received. The result ofthe fuzzy rules will be the composite event, its fuzzy value, and the degree of membership. The fuzzy value showshow serious is the detected event and the degree of membership shows the likelihood of the detected compositeevent. Higher membership degrees indicate that the CE has happened with higher certainty and lower values, viceversa.

4.4. Detection process

The event detection process consists of three main activities, which are *simple event detection*, *composite event detection*, and *event stream maintenance*.

Ordinary sensor nodes are responsible for detecting simple events. The activity starts when a sensor node detectsan event (*event name*) based on the sensed values and predefined threshold values for the simple events. In the nextstep, the node maps the sensed value to a fuzzy one (*fuzzy value*) using the fuzzy membership function, andcalculate the corresponding membership degree. Finally in the last step, the complementary information such as*node ID* and *detection timestamp* are added to the event notification and submitted to the fusion point.

The coordinator is responsible for composite event detection based on the received event notifications from itscluster members. The process triggering mechanism is a design issue aspects of the process. The process of detectingcomposite events can be triggered either by arrival of new event notification, or by a timer. The former providesfaster detection with the price of increased processor consumption, while the detection speed in the latter is sensitiveto the timer value. The occurrence frequency of an event, which is an application specific parameter, is one of thefactors that play an important role in choosing either case.

One of the other responsibilities of the coordinator is to analyze the correlations between the received notifications.In *FED* the event correlation is investigated from two different aspects. The first parameter is the time distance(*detection window*) between the reported simple events and the second one is the physical distance between thereporting nodes. The output of the correlation analysis step is a set of event notifications that should be examined touncover a possible composite event. Then the correlated notifications are aggregated based on the methods describedin Section 4.3.

Maintenance of the received event streams at the fusion point helps to improve the efficiency of the model. Oneof the main responsibilities of cluster heads in *FED* is to keep track of the correlated simple events. Therefore, thereported event stream has to be scanned frequently by the cluster head. It is highly recommended to keep the storedstream as short as possible to save memory and processor resources. One of the ways to keep the stream shortis to set an expiration time for the stored events. Consequently, the expired event notifications will be removedautomatically. Two parameters should be considered in defining the expiration time effectively. The first parameteris the largest detection window for the composite event and the second one is the time synchronization accuracy[36].

4.5. Uncertainty

Generally, a predefined set of simple events should happen in order to be able to detect a composite event. Thereare cases where at least one of the required simple events misses, due to various reasons. To cope with the problem,*FED* calculates the occurrence likelihood of the composite event. The calculation is similar to aggregation of simpleevents. The output of the calculation is a membership degree of a possible composite event in the absence of arequired simple event. *FED* uses threshold called acceptance ratio, to recognize the reported simple events as acomposite event. Lower values for acceptance ratio threshold will result on higher false positives while highervalues will disable the uncertainty handling of *FED*.

For example consider the fuzzy rule of Section 4.3 and the following event notifications from the two nearbysensor nodes.

$$y_1 = (heat, 12, 12:23: 40 \text{ GMT}, medium, 0.90) \tag{7}$$

$$y_2 = (smoke, 20, 12:23: 39 \text{ GMT}, medium, 0.85) \tag{8}$$

In the absence of the third simple event, *FED* calculates the average of the membership degrees which is 0.583. Ifthe acceptance ratio for the rule is below the calculated number, the composite event will be detected.

5. EVALUATION

To evaluate the performance of the introduced model on detecting composite events under uncertainty sources,several of simulation configurations were setup. Before analyzing the achieved results, we would like to introducethe simulation tool first.

5.1. Simulation environment

We have developed a simulation environment based a software agent development tool called JADE [37]. *JADE*is a framework for developing multi-agent systems. These are some of the advantages of using *JADE* for simulatingwireless sensor network algorithms.

- The framework encapsulates the network protocol stack and helps to ignore the hardware level details of sensor nodes. Considering each node as a software agent gives us a chance to analyse the higher level behaviour of the algorithm in the network.

- The autonomous property of software agents maps well with sensor nodes' autonomous nature.

- Message passing is considered as the only communication mechanism in *JADE* and wireless sensor networks. That is, there are no method calls or shared memory facilities in both cases. Arrived messages are stored in a FIFO queue in each agent in *JADE* which is similar to a sensor node. Thisalso provides the means of having an asynchronous communication.

The aforementioned features give us sufficient reasons to build our simulation based on *JADE*. But in order to fitthe tool with the problem criteria, we have added five additional features.

1) Event notifications omitted randomly to simulate message loss in the network.

2) To simulate false detection behavior of nodes, a random false detection mechanism is added to sensor nodes.

3) We have added a delay in message transmission in order to simulate propagation delay using a Gaussianrandom generator.

4) Sensor nodes are only allowed to communicate with each other, only if they are within each other's transmission range. All the sensor nodes can communicate with the coordinator in a bidirectional way.

5) An event generator creates CEs by defining the exact location randomly and sends the message to only thosenodes that are within the sensing rage of the generated event.

5.2. Simulation results

To evaluate *FED*, we simulated an experimental case similar to explosion detection application. In our simulation,heterogeneous nodes are distributed in the environment uniformly random. We use three types of sensors(SensorType1, SensorType2 and SensorType3) with equal quantities. In each experiment a sensor field is generatedwith predefined number of nodes that has been located uniformly distributed. 100 CEs with random locations aregenerated periodically across the network area of $600 * 600$ unit^2. To reduce the effect of the random distributionwe have run each experiment 50 times and averaged the results.

In the first set of experiments, we investigate the effect of sensing coverage area of a sensor and the message lossrate on the percentage of detected composite events. Figure 2 shows the simulation results for a network consistsof 12 nodes from 3 sensor types (4 sensors from each type) which are deployed randomly. Besides, false simple event reports are produced randomly with rate of 5%. The evaluation metric (Y axis) is the percentage of detectedcomposite events. The performance of *FED* is compared with the approach introduced in [27]. According to theexperiment scenario, we increase the sensing range of sensor nodes from 240 to 380 units. The lower sensing rangerepresents less dense network, while the higher sensing ranges show the dense ones. The reason is, in the highersensing ranges each point in the network field may be covered by more nodes, whereas in lower sensing rangesit will be covered by fewer nodes. The first thing that the simulation results show is that, increasing the sensingrange will provide higher detection percentage of composite events.

To investigate the effect of message loss, we apply two different message loss rates. In the first case 25% ofsimple event notifications are omitted randomly. The results show that *FED* outperforms 15% with the sensingrange of 240 units. As the sensing range increases, both approaches achieve higher percentage of event detection.

The results also show that *FED* is less sensitive to message loss. That is, the difference between the percentagesof detected events in the lowest and highest sensing ranges for the case of 25% of message loss is 35% for *FED*and 42% for [27]. The results for the case of 15% of message loss also support the idea.

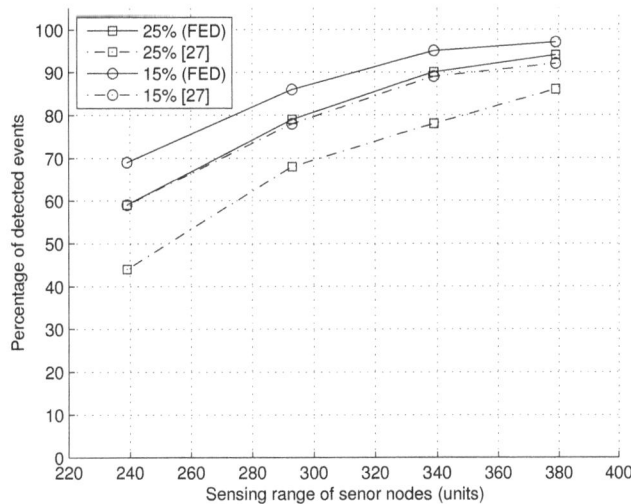

Figure2. CE detection in a network with node density of 3 false alarm 5%.

We run the simulation with similar configuration with 24 nodes. Figure 3 shows the simulation results, whereX axis is the sensing range and the Y axis is the percentage of detected composite events. The results show that*FED* performs better in all the sensing ranges. For instance, *FED* detects 57% of composite events for the casewhere sensing range is 169, while the approach in [27] detects only 43% in the presence of 25% message loss.The difference between the detection percentages of both approaches reduces in the dense network, due to multipledetection of a simple event. But even in the sensing range of 268, *FED* detects 6% more composite events. Moreover, similar to previous experiment, the results show that *FED* is less sensitive to environmental noise. For example,the difference between the highest and lowest percentage of event detection is 37% and 46% for *FED* and [27]respectively with 25% of message loss. In the network with 15% of message loss, the difference is 30% and 40%for *FED* and [27] respectively.

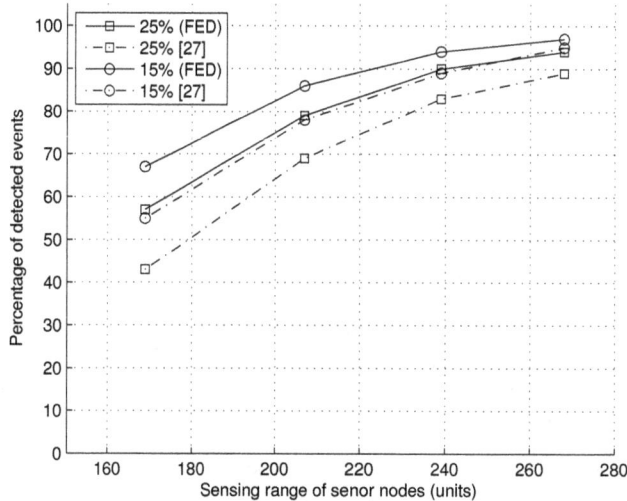

Figure3. CE detection in a network with node density of 3 false alarm 5%.

Figure 4 and Figure 5 shows the experimental results for the network of 36 and 48 nodes respectively. We alsoincrease the false alarm rate to 10%. The outcome of the experiments is similar to the previous analysis.

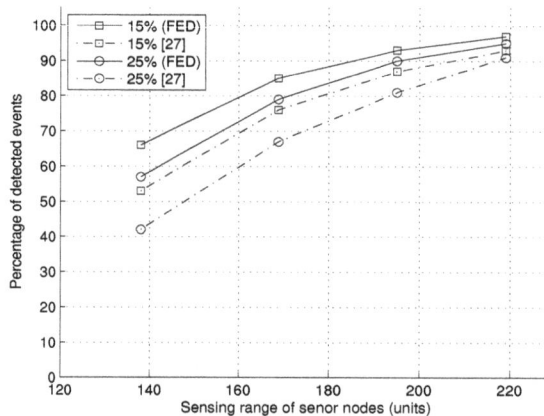

Figure4. CE detection in a network with node density of 4 false alarm 10%.

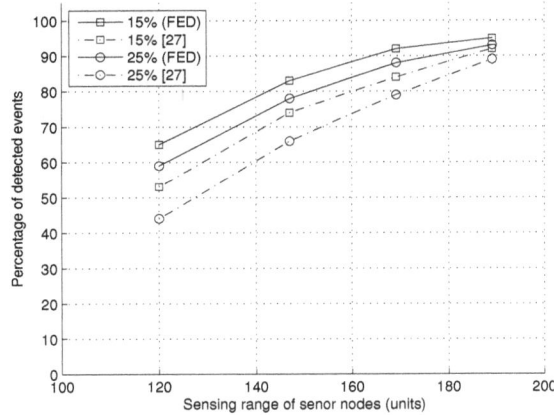

Figure5. CE detection in a network with node density of 4 false alarm 10%.

In the next set of experiments we investigate the effect of message loss in detail. The network surface is the sameas previous experiments and the quantity of each sensor type is equal to $\frac{1}{3}$of network nodes. Figure 6 demonstratesthe percentage of detected events (Y axis) for various message loss rates (X axis). The graph shows the results forthe network of 12 and 24 nodes and false alarm rate is set to be 10%. The sensing range is defined in way that thesenetwork configurations have the same node density. The sensing range for the network of 12 nodes is set to be 239units and for the case of a 24 node network is 169 units. To calculate the proper sensing range we use the followingequation where d, L, W, and n are network node density, length, width, and size (number of nodes) respectively.

$$t_r = \sqrt{\frac{dLW}{2\pi n}} \qquad\qquad (9)$$

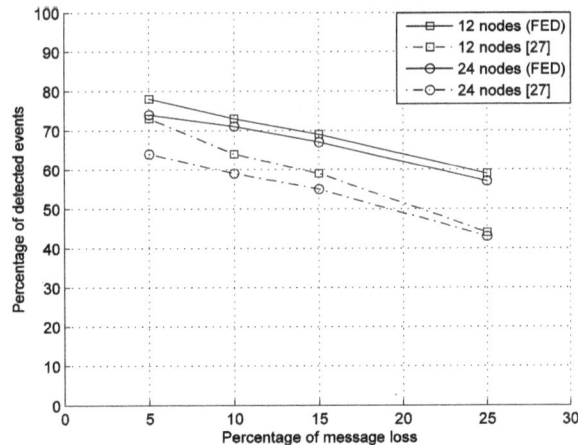

Figure6. *FED* sensitivity to quantity of nodes under 10% false alarm rate.

Similar to the previous results, *FED* detects more events compared to the approach introduced in [27]. For examplein a network of 12 with 5% message loss, *FED* detects 5% of events that is not detected by [27]. As the messageloss rate increase, the percentage of detected events in both approaches decrease. But the results show that *FED* is more robust message loss. For the case of 25% message loss in a network of 12 nodes, *FED* uncovers more than15% of undetected events by [27], while in the presence of 5% message loss was 5%.

Figure 7 depicts the results for a similar experiment with 36 and 48 nodes. The node density is 3 which means that the sensing range for each node in the 36 node network is 169 units and for the network of 48 nodes is 148units and the false alarm rate is 10%.

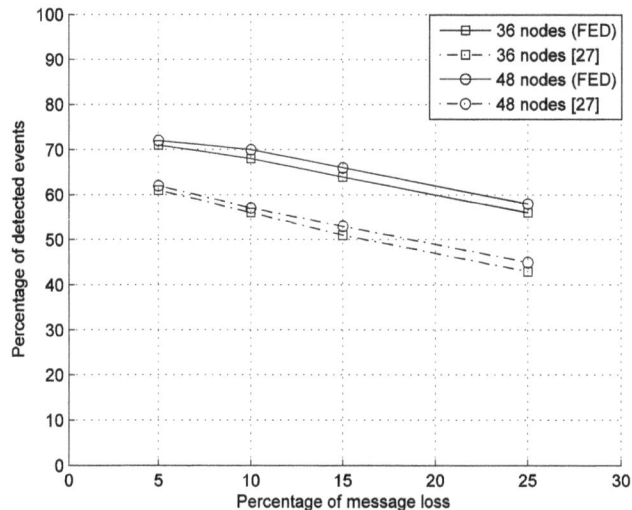

Figure7. *FED* sensitivity to quantity of nodes under 5% false alarm rate.

6. Conclusions

In this paper we have introduced a fuzzy model called *FED* for distributed detection of composite events. *FED* supports the heterogeneity of sensor types. To perform distributed detection, the model uses a diversity based clustering scheme in which each cluster considers maximizing the sensor diversity of their member nodes.Therefore, the clusters are able to detect a wide range of composite events. Besides, the model provides node level knowledge abstraction which is a valuable characteristic in designing heterogeneous sensor networks. A fuzzy approach is used to uncover composite events in the presence of uncertainty sources such as message loss and false alarms. The simulation results show that *FED* outperforms in terms of the percentage of detected events. The amount of improvement is significant in networks with low node density. Besides, *FED* is suitable for the networks with high message loss rates.

Acknowledgements

This research is funded by Iranian Telecommunication Manufacturing Company (ITMC).

REFERENCES

[1]   M. Ceriotti, L. Mottola, G. P. Picco, A. L. Murphy, S. Guna, M. Corra, M. Pozzi, D. Zonta, and P. Zanon, "Monitoring heritage buildings withwireless sensor networks: The torreaquila deployment," in Proceedings of the 2009 International Conference on Information Processingin Sensor Networks, ser. IPSN '09. Washington, DC, USA: IEEE Computer Society, 2009, pp. 277–288.

[2]   R. Szewczyk, A. Mainwaring, J. Polastre, J. Anderson, and D. Culler, "An analysis of a large scale habitat monitoring application," inProceedings of the 2nd international conference on Embedded networked sensor systems, ser. SenSys '04. New York, NY, USA: ACM,2004, pp. 214–226.

[3]   K. Martinez, J. K. Hart, and R. Ong, "Environmental sensor networks," Computer, vol. 37, pp. 50–56, 2004.

[4]   A. Yilmaz, O. Javed, and M. Shah, "Object tracking: A survey," ACM Comput. Surv., vol. 38, December 2006.

[5]   E. I. Shih, A. H. Shoeb, and J. V. Guttag, "Sensor selection for energy-efficient ambulatory medical monitoring," in Proceedings of the7th international conference on Mobile systems, applications, and services, ser. MobiSys '09. New York, NY, USA: ACM, 2009, pp.347–358.

[6]   M. F. Duarte and Y. H. Hu, "Vehicle classification in distributed sensor networks," J. Parallel Distrib.Comput., vol. 64, pp. 826–838, July2004.

[7]   M. Keally, G. Zhou, and G. Xing, "Watchdog: Confident event detection in heterogeneous sensor networks," Real-Time and EmbeddedTechnology and Applications Symposium, IEEE, vol. 0, pp. 279–288, 2010.

[8]   T. He, S. Krishnamurthy, L. Luo, T. Yan, L. Gu, R. Stoleru, G. Zhou, Q. Cao, P. Vicaire, J. A. Stankovic, T. F. Abdelzaher, J. Hui, andB. Krogh, "Vigilnet: An integrated sensor network system for energy-efficient surveillance," ACM Trans. Sen. Netw., vol. 2, pp. 1–38,Feb. 2006.

[9]   S. R. Madden, M. J. Franklin, J. M. Hellerstein, and W. Hong, "Tinydb: an acquisitional query processing system for sensor networks,"ACM Trans. Database Syst., vol. 30, pp. 122–173, March 2005.

[10]   T. Liu and M. Martonosi, "Impala: a middleware system for managing autonomic, parallel sensor systems," SIGPLAN Not., vol. 38, pp.107–118, June 2003.

[11]   W. B. Heinzelman, A. L. Murphy, H. S. Carvalho, and M. A. Perillo, "Middleware to support sensor network applications," IEEE Network,vol. 18, no. 1, pp. 6–14, 2004.

[12]   E. Souto, G. Guimar˜aes, G. Vasconcelos, M. Vieira, N. Rosa, and C. Ferraz, "A message-oriented middleware for sensor networks," inProceedings of the 2nd workshop on Middleware for pervasive and ad-hoc computing, ser. MPAC '04. New York, NY, USA: ACM,2004, pp. 127–134.

[13]   T. Abdelzaher, B. Blum, Q. Cao, Y. Chen, D. Evans, J. George, S. George, L. Gu, T. He, S. Krishnamurthy, L. Luo, S. Son, J. Stankovic,R. Stoleru, and A. Wood, "Envirotrack: Towards an environmental computing paradigm for distributed sensor networks," DistributedComputing Systems, International Conference on, vol. 0, pp. 582–589, 2004.

[14]   R. Kumar, M. Wolenetz, B. Agarwalla, J. Shin, P. Hutto, A. Paul, and U. Ramachandran, "Dfuse: a framework for distributed data fusion,"in Proceedings of the 1st international conference on Embedded networked sensor systems, ser. SenSys '03. New York, NY, USA: ACM,2003, pp. 114–125.

[15]   S. Li, S. H. Son, and J. A. Stankovic, "Event detection services using data service middleware in distributed sensor networks," in Proceedingsof the 2nd international conference on Information processing in sensor networks, ser. IPSN'03. Berlin, Heidelberg: Springer-Verlag,2003, pp. 502–517.

[16]   J. Mao, J. Jannotti, M. Akdere, and U. Cetintemel, "Event-based constraints for sensornet programming," in Proceedings of the secondinternational conference on Distributed event-based systems, ser. DEBS '08. New York, NY, USA: ACM, 2008, pp. 103–113.

[17]   R. Rajkumar, I. Lee, L. Sha, and J. Stankovic, "Cyber-physical systems: The next computing revolution," in Design Automation Conference(DAC), 2010 47th ACM/IEEE, june 2010, pp. 731 –736.

[18]   S. Ren, "Apeser 2010 keynote speech: Shangpingren," in Green Computing and Communications (GreenCom), 2010 IEEE/ACM Int'lConference on Int'l Conference on Cyber, Physical and Social Computing (CPSCom), dec. 2010, p. lvix.

[19]   E. Lee, "Cyber physical systems: Design challenges," in Object Oriented Real-Time Distributed Computing (ISORC), 2008 11th IEEEInternational Symposium on, may 2008, pp. 363 –369.

[20]   M. Hill, M. Campbell, Y.-C. Chang, and V. Iyengar, "Event detection in sensor networks for modern oil fields," in Proceedings of thesecond international conference on Distributed event-based systems, ser. DEBS '08. New York, NY, USA: ACM, 2008, pp. 95–102.

[21]   V. N. S. Samarasooriya and P. K. Varshney, "A fuzzy modeling approach to decision fusion under uncertainty," Fuzzy Sets Syst., vol. 114,pp. 59–69, August 2000.

[22]   R. Viswanathan and P. Varshney, "Distributed detection with multiple sensors i. fundamentals," Proceedings of the IEEE, vol. 85, no. 1,pp. 54 –63, Jan. 1997.

[23]   V. Saligrama, M. Alanyali, and O. Savas, "Distributed detection in sensor networks with packet losses and finite capacity links," SignalProcessing, IEEE Transactions on, vol. 54, no. 11, pp. 4118 –4132, 2006.

[24]   E.Ermis and V.Saligrama, "Distributed detection in sensor networks with limited range multimodal sensors," Signal Processing, IEEETransactions on, vol. 58, no. 2, pp. 843 –858, 2010.

[25]   H. TabatabaeeMalazi, A. Khalili, K. Zamanifar, and S. Dulman, "DEC: Diversity based energy aware clustering for heterogeneous sensornetworks," Ad Hoc and Sensor Wireless Networks (Accepted).

[26]   H. TabatabaeeMalazi, K. Zamanifar, A. Pruteanu, and S. Dulman, "Gossip-based density estimation in dynamic heterogeneous sensornetworks," in Wireless Communications and Mobile Computing Conference (IWCMC), 2011 7th International, July 2011, pp. 1365 –1370.

[27]   P. Manjunatha, A. Verma, and A. Srividya, "Multi-sensor data fusion in cluster based wireless sensor networks using fuzzy logic method,"in Industrial and Information Systems, 2008.ICIIS 2008.IEEE Region 10 and the Third international Conference on, 2008, pp. 1 –6.

[28]   A. V. U. Phani Kumar, A. M. Reddy V, and D. Janakiram, "Distributed collaboration for event detection in wireless sensor networks,"in Proceedings of the 3rd international workshop on Middleware for pervasive and ad-hoc computing, ser. MPAC '05. New York, NY,USA: ACM, 2005, pp. 1–8.

[29]   H. Chan and A. Perrig, "ACE: An emergent algorithm for highly uniform cluster formation," in European Workshop on Wireless SensorNetworks (EWSN 2004), Jan. 2004.

[30]   S. Soro and W. B. Heinzelman, "Cluster head election techniques for coverage preservation in wireless sensor networks," Ad Hoc Netw.,vol. 7, pp. 955–972, July 2009.

[31]   O. Younis and S. Fahmy, "Distributed clustering in ad-hoc sensor networks: a hybrid, energy-efficient approach," in INFOCOM 2004.Twenty-third AnnualJoint Conference of the IEEE Computer and Communications Societies, vol. 1, 2004, pp. 4 vol. (xxxv+2866).

[32]   S. K. Singh, M. P. Singh, and D. K. Singh, "Energy efficient homogenous clustering algorithm for wireless sensor networks," International Journal of Wireless & Mobile Networks (IJWMN), vol.2, no.3, August 2010.

[33]   Getsy S Sara, Kalaiarasi.R, NeelavathyPari.S and Sridharan .D, "Energy efficient clustering and routing in mobile wireless sensor network," International Journal of Wireless & Mobile Networks (IJWMN) vol.2, no.4, November 2010.

[34]   S.Taruna, Kusum Jain and G.N. Purohit, "Power Efficient Clustering Protocol (PECP)-Heterogeneous Wireless Sensor Network," International Journal of Wireless & Mobile Networks (IJWMN) vol.3, no.3, June 2011.

[35]   L. Zadeh, "Fuzzy sets," Information and Control, vol. 8, no. 3, pp. 338 – 353, 1965.

[36]   B. Sundararaman, U. Buy, and A. D. Kshemkalyani, "Clock synchronization for wireless sensor networks: a survey," Ad Hoc Networks,vol. 3, no. 3, pp. 281 – 323, 2005.

[37]   Java agent development framework @ONLINE. [Online]. Available: http://jade.tilab.com/

10

ASRoP : Ad hoc Secure Routing Protocol

Rida Khatoun, Lyes Khoukhi, Ahmed Nabet and Dominique Gaïti

ICD - ERA - University of Technologies of Troyes (UTT), STMR, UMR CNRS 6279
12, rue Marie Curie   10000 - Troyes, France

Email: rida.khatoun@utt.fr

Abstract

Mobile ad hoc networks (MANETs) are a new concept of wireless communications for mobile devices, which offer communications over a shared wireless channel without any pre-existing infrastructure. Their wireless nature and self-organizing capabilities are some of MANET's biggest advantages, as well as their biggest security restrictions. Forming end-to-end secure paths in such MANETs is more challenging than in conventional wireless cellular/wired networks due to the lack of central authorities. An attacker can easily disrupt the routing process by injecting false control messages, changing the paths of packets or simply by blocking the packets of other nodes. In this paper, we propose a novel efficient secure routing protocol, named ASRoP, to effectively secure the routing discovery process in ad hoc networks. ASRoP provides powerful security extensions to the reactive AODV protocol, based on Diffie-Hellman (DH) algorithms and our modified secure remote password protocol. The simulation results show the efficiency of the proposed ASRoP protocol, and its cost towards both the users and the network. ASRoP promises to offer a real opportunity to prevent attacks related to lack of authentication without degrading routing performance.

Keywords

Wireless Network, Ad hoc Network, Security, Wireless Routing Protocol

1. Introduction

In recent years, wireless ad hoc networks (MANETs) have received tremendous attention because of their self-maintenance and self-configuration capabilities. A MANET is a set of autonomous wireless mobile devices that communicate with each other over wireless links. Such networks do not require the deployment of any infrastructure for their operation; thus, it is expected that they will play a vital role in future civilian and military settings, being useful to provide communication support where the deployment of a fixed infrastructure is not possible or economically profitable. The topology of MANETs is in general dynamic, because the connectivity among the nodes may change with time due to the dynamics of nodes or churn. Communication is performed by relaying data packets along suitable routes, which are dynamically discovered and maintained through cooperation between the nodes; thus, any routing protocol design must consider the limitations and constraints of MANETs.

Several routing schemes have been proposed in the literature (e.g. AoDV [1], DSR [2], etc.). These schemes focus mainly on finding routes between sources and destinations nodes, and on efficiency issues such as scalability with respect to network size and traffic load [3]. Usually, the length of routes is the main metric used in these schemes. It is observed that most of these routing schemes have ignored the aspect of network security; thus, they are vulnerable to attacks since they do not consider a secure path during the route discovery process. To alleviate this limitation, several approaches have been proposed to secure ad hoc routing. Some of these approaches employ mechanisms used to protect routing protocols in wired networks based on the presence of a centralized infrastructure; however, these solutions may not be appropriate for a decentralized environment such as ad hoc network.

In this paper, we propose a novel secure routing scheme for mobile ad hoc networks based on AODV on-demand protocol, named ASRoP. It is specifically designed to an open wireless ad hoc network where each node should verify the identity of the node with which it communicates. On the contrary to the classical use of remote secure routing protocol which is actually employed in "client-server" context; our contribution proposes to use a modified version of this protocol in distributed ad hoc mobile environment. This allows nodes to be authenticated before considering any information during the route discovery phase. In our ASRoP, we focus on attacks carried out by traditional external illegitimate nodes which do not have the access rights to the ad hoc network. Our protocol considers also some other attacks that may be carried out by internal malicious nodes that inject false information about the network topology. Moreover, the proposed protocol ensures the reliability of the route (s) obtained during the route discovery phase. The contributions of this paper can be summarized as follows:

- a new security protocol based on AODV that ensures the establishment of a secured routes between source and destination nodes, while it reduces the load of cryptographic functions conventionally used

- a new way of detecting and rejecting forged or replayed messages

- a new key exchange method achieved in a fully distributed fashion without any need for a permanent or temporary infrastructure

The rest of the paper is organized as follows. Section 2 highlights some vulnerabilities of MANETs and presents a brief state of the art of some secure routing solutions. In Section 3, we detail our proposed secure routing protocol ASRoP. The performance evaluation of ASRoP are presented in Section 4. Finally, Section 5 concludes this paper.

This document describes, and is written to conform to, author guidelines for the journals of AIRCC series. It is prepared in Microsoft Word as a .doc document. Although other means of preparation are acceptable, final, camera-ready versions must conform to this layout. Microsoft Word terminology is used where appropriate in this document. Although formatting instructions may often appear daunting, the simplest approach is to use this template and insert headings and text into it as appropriate.

2. ROUTING SECURITY IN AD HOC NETWORKS

2.1 Vulnerabilities in ad hoc networks

Despite the fact that wireless networks are more flexible than wired networks, they are also more vulnerable to attacks. This is due essentially to the fact that wireless channel is accessible to both legitimate network users and malicious attackers. In traditional wired networks; to eavesdropping, an intruder would need to listen physically to the cable. However, in wireless networks, an attacker is able to listen to all messages in the transmission area [4]. Therefore, just by being in the same coverage area, the intruder has access to network communications and can easily intercept the data transmitted without the sender knowing about the intruder attacks. As the intruder is virtually invisible, it can also record, edit, and then retransmit packets as they are issued by the sender, even claiming that the packets originate from a legitimate party. In addition, due to environmental constraints, wireless communications can easily be disturbed; the attacker can perform this attack by generating some noises. Attacks on MANET could be categorized in 5 layers; Application layer, Transport layer, Network layer, Link layer, and Physical layer [5. In this paper, we focus on network layer attacks; this kind of attacks aims to change the routing protocol to redirect traffic to a specific node that is under the control of attackers. An attack may also prevent the construction of the network, announcing incorrect routes, and more generally to corrupt the network topology [6][7]. Routing attacks can be classified into two categories: incorrect traffic generation and incorrect traffic relay.

2.1.1 Incorrect traffic generation

This category includes attacks that involve sending false messages. For example, control messages sent on behalf of another node (spoofing), or control messages containing incorrect or outdated routing information. The network can present a Byzantine behavior [8] [9], i.e. contradictory information sent from different parts of the network. The consequences of this attack are the degradation in network communications, isolated nodes and routing loops [10] [11]. Cache Poisoning and DoS (Denial of Service) are examples of incorrect traffic generation in routing schemes.

- Cache Poisoning: in the distance vector routing protocol, an example of incorrect traffic generation is that an attacker may announce a metric of 0 for all destinations, which will induce all nodes send their packets through this node. Then the node deletes all packets which will cause a significant loss of exchanged communications.

- DoS: an attacker can perform a denial of service in the network by saturating the wireless medium with broadcast messages, which will reduce the rate of transmission of nodes and prevent communications. An attacker can send invalid messages just to paralyze nodes, overload their CPU and consume their energy resource. In this case, the attack aims not to change the network topology, but rather to disrupt network functionality and communications.

2.1.2 Incorrect traffic routing

Information sent from a legitimate node can be corrupted by another node [12] [13]. Examples of this category are:

- Black hole attack: malicious nodes falsely claim a fresh route to the destination to absorb transmitted packets from source to that destination and drop them instead of forwarding.

- Green Hole: the attacker distributes a portion of the received messages and blocks the others. For example, it filters the data packets to be hidden and passes the control packets.

- Message tampering: an attacker can also modify messages before forwarding them. This is may be happen only if no mechanism is used to ensure the integrity of data packets.

- Replay attack: as the topology is dynamic, an attacker can produce replay attacks, using control messages already recorded and transferred to other nodes in order to modify the nodes routing tables with false information.

- Rushing attack: this attack can be launched against reactive protocols. In these protocols, nodes only rebroadcast the first request received for each route discovery and ignore others. When a route discovery is initiated, the attacker floods the network by request messages. If the attacker's messages arrive firstly, the attacker will be involved in the route discovery process [14].

- Wormhole Attack: two malicious nodes cooperate and falsify the number of hops by announcing a short cut between two imaginary parts of the network. In the Figure 1, the source S chooses to route the data packets by {S, M1, M2, D} instead of {S, A, B, C, D} because it is the shorter route but in reality, attackers use a longer route {S, M1, A, B, C, M2, D} since that the link between M1 and M2 is unreal.}

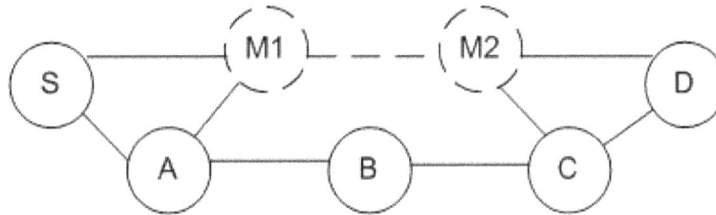

Figure 1. Example of Wormhole attack

According to the authors of [12] [15] [10] [6] [16], attacks are classified according to several criteria:

- internal or external attack: if the attacker compromises an existing node, the attack is considered as internal. Otherwise, it is an external attack.

- active or passive attack: the purpose of an active attack is to modify the protocol so that traffic packets passes through a node which is controlled by an attacker. A passive attack takes control over a corrupted node to eavesdrop the traffic. This does not jeopardize the functionality of the network, but affects the anonymity of the exchanged messages. This attack allows the attacker to analyse data packets that can be used later.

- single or distributed attack: In a single attack, a single entity is used. More sophisticated attacks, called distributed attacks, compromise several nodes and are generated from various sources. This kind of attack is more dangerous and difficult to detect.

2.2. Routing security in MANETs

The traditional routing protocols aim only to discover optimal routes. Some current works have introduced the concepts of security in these protocols without changing their basic principles by including authenticity and integrity of the exchanged messages. Some research works have targeted a specific protocol and present themselves as an extension of its original version, or they provide some countermeasures against attacks. In what follows, we summarize some relevant protocols which are classified into several categories according to the mechanism used.

2.2.1 Solutions based on asymmetric cryptography

The protocols using asymmetric cryptography need a trusted third party or so-called *TTC* which provides certificates to the participating nodes. ARAN and SAODV are examples of these protocols.

- ARAN: Authenticated Routing for Ad-Hoc Networks [17]. This protocol uses a reactive trusted party to generate certificates. Before joining the network, each node must obtain a certificate signed by the server. In ARAN, each node signs the discovery and replay packets before retransmitting. Intermediate nodes cannot reply with a route replay. Only the destination node has the right to respond. This provides an end-to-end authentication between the source and destination; however this leads to increase dramatically the latency especially for long routes.

- SAODV: Secure Ad hoc On-demand Distance Vector Routing [18]. It is a secured extension of AODV protocol. In this protocol, the non-mutable fields in AODV messages are signed by the private key of the sender while a hash function is used to protect the fields' integrity. The signature ensures the functions of authentication, integrity and non-repudiation. Despite its robustness, SAODV is vulnerable to the wormhole attack.

2.2.2 Solutions based on symmetric cryptography

This set of solutions is built on symmetric cryptography, hash functions, and hash chains.

- SRP: Secure Routing Protocol [18]. It was designed to provide trust routing information by securing the route discovery step. SRP requires private keys shared by the hosts. The destination checks the integrity and authenticity of routing messages using a hash function then broadcasts its reply by different routes. This later technique permits an additional protection against malicious nodes that attempt to alter *route replay* messages. The weakness of this protocol is at the route discovery: an adversary can produce *Route Error message* to invalidate routes that are still available.

- SEAD: Secure Efficient Ad hoc Distance vector routing protocol [19]. It is a proactive protocol developed for securing DSDV protocol. SEAD authenticates the sender and provides protection against the tampering of mutable fields (e.g., the number of hops, sequence number). By applying a hash function repeatedly on a random value, a chain of hash is obtained. Then, the elements of this chain are used by the nodes in the authentication procedure without a need for public key encryption. This avoids the costly cryptographic operations. To authenticate the source of an update, a shared secret key between each pair of node is required.

2.2.3 Solutions based on reputation

This solution addresses the selfish behavior problem which considerably disrupts the routing process. The main goal of a reputation system is to make decisions about the reliability of entities and improve trust within the network by encouraging the participation in routing. To make such decisions, a reputation system analyzes ancient interactions and exchanges between nodes. Each scheme discussed above has its own requirements and constraints to achieve the desired security. Protocols based on the cryptographic mechanism require key management. Protocols based on reputation include a new metric (reliability of the path) to select a route to a destination. Intermediate nodes are limited to route packets in some protocols while in others they are permitted to respond to the source if they know the path. While many theoretical studies have been proposed in the literature, the satisfaction of safety constraints inherent in ad hoc infrastructure still need more investigation.

3. PROPOSED SOLUTION: AD HOC SECURE ROUTING PROTOCOL (ASRoP)

Our proposed Protocol ASRoP is inspired from SRP authentication algorithm [20]; this is due mainly to the effectiveness of the hash chains used in SRP since they reduce the costs of the traditional cryptographic mechanisms. We note that almost methods based on authentication via certificates are very expensive and may be not enough secured if a certificate was not issued by trusted third-party. Since ad hoc networks are very dynamic, the certificate-based solutions may be limited. It is important to note that a password in SRP is never transmitted during the messages exchange process, even encrypted. In this way, a hacker node cannot intercept the password by listening to the network; and this prevents problems of using passwords for authentication.

Our ASRoP solution includes two main steps: in the first one the nodes exchange information to negotiate the parameters which are required to establish a shared secret information between neighbors; this secret information is used as a password in the second step. In the second step, the nodes involved in the route discovery verify the identity of each node that provides information about the route. Note that in ASRoP, *HELLO* packets are sent at regular intervals to update the neighboring table which permits to any node to obtain a global view about its neighborhood.

3.1. Authentication scheme

As we stated previously, our proposal is based on SRP [20] which is a secure password-based authentication and key-exchange protocol; it is efficient for negotiating secure connections using a password provided to users, while eliminating the security issues typically associated with reusable passwords. This protocol also performs a secure key exchange in the process of authentication, allowing security layers (privacy protection and/or integrity) to be activated during the session. Key servers and certificate infrastructures are not required. The following steps, illustrated in Figure 2 explain the process of authentication *Client/Server* or two nodes in general (we included some modifications in the results verification phase).

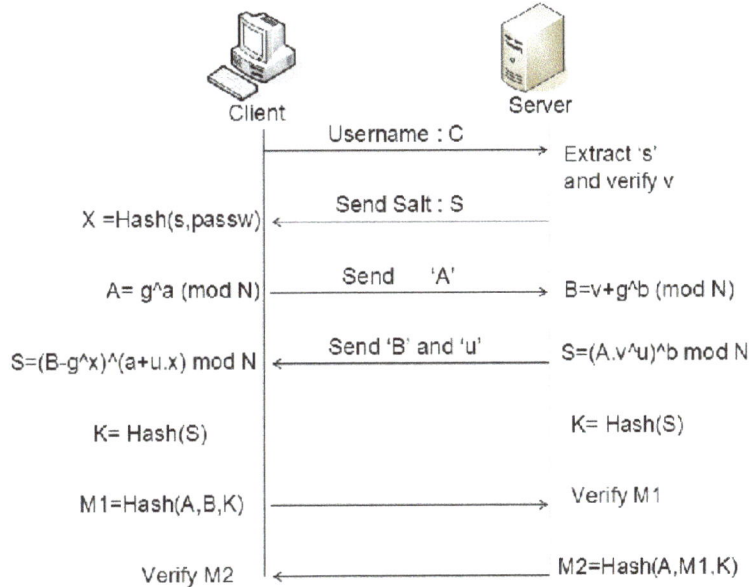

Figure 2. Secure remote password protocol authentication

1. The two entities choose two distinct numbers g and N. N is a prime number and g is a primitive root modulo N.

2. The client calculates $x=hash(s \| hash(UserName \| ":" \| Password))$ where s is a random string used as the user's salt. Salt is known by both user and server.

3. The client sends its username to the server.

4. The server checks the password entry, calculates a verifier $v=g^x \bmod N$.

5. The client generates a random number a which is a private key generated randomly and not publicly revealed, calculates A and sends A to the server.

6. The server generates its own random number b which is a private key generated randomly and not publicly revealed, calculates B and generates another random parameter u and sends it with B to client. u is random scrambling parameter obtained from B (the MSB 32 bits of hash(B)).

7. Both Client and server compute the same value S using the available values but with different operations.

8. If the password P of client entered in step 2 corresponds to that used in the calculation of v, the two S will match.

9. Both entities hash *S* to create a session key cryptographically strong.

10. The client sends *M1* to the server as proof that it has the right session key. The server computes *M1* and verifies that it corresponds to that sent by the client.

11. The server sends *M2* to the client as proof that it has the right session key. The client computes *M2* and verifies that it corresponds to that sent by the server.

In ASRoP, the authentication process is performed during the route discovery phase, and specifically when a destination node sends response message to a source node. During this step, each node involved in the creation of the route does not exploit any information from a node only if this later is authenticated. The authentication is ensured by the modified SRP protocol using the shared key obtained in the first step. Upon receiving the route request, the intermediate node has two possibilities: either it responds to the source if it is the destination or it has a valid route to reach the destination. In the first case, the intermediate node initiates an authentication process with the next hop. The authentication is performed hop by hop along the path.

3.2. Keys exchange

At this stage, all nodes exchange information to establish a secret key with their neighbors (at one hop) by applying the *Diffie-Hellman (DH)* algorithm. Each node sends a request for each neighbor in order to share a secret, and each neighbor responds to the request, using parameters generated according to DH algorithm. Figure 3 shows the step dealing with a secret sharing between nodes.

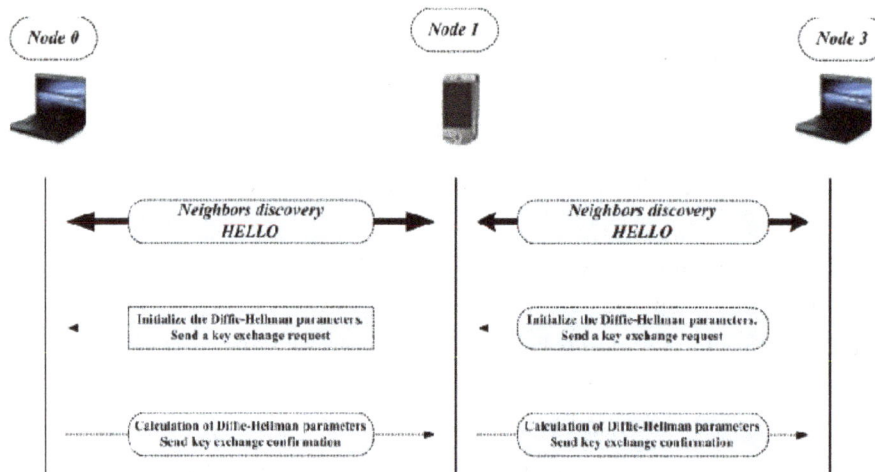

Figure 3. Exchange keys between nodes

Upon receiving a Hello message that reveals a neighbor, the node decides whether to reply or not, the decision is taken based on the identifier (ID) of the neighbor. Only the nodes having an ID greater than their neighbors can initiate a key exchange request (Figure 4). Following this step, each pair of nodes shares a secret which can only be known by these nodes; thus, all nodes share secret keys with their neighbors.

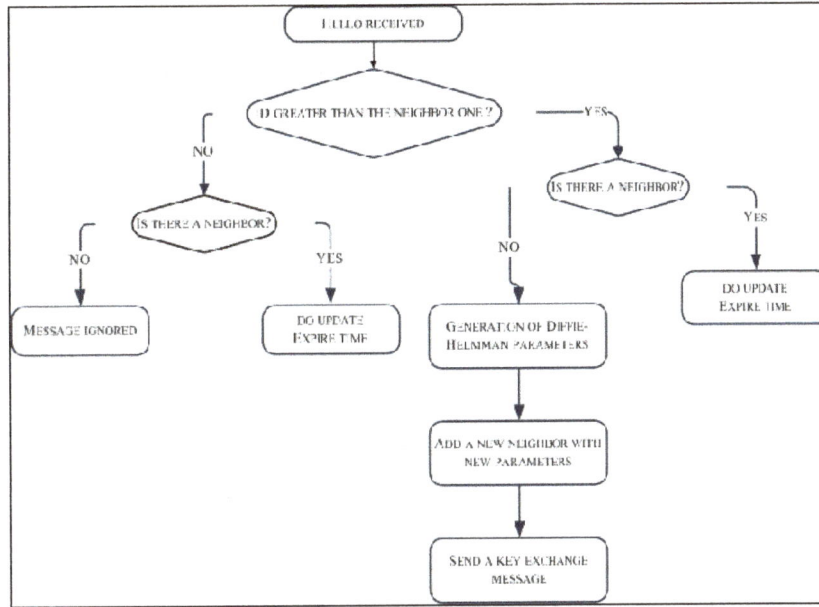

Figure 4. Decision policy of Hello packet reception

3.3. Format of messages in ASRoP

The design of our proposed ASRoP is based on AODV routing protocol, which is an efficient reactive routing scheme. To adapt AODV to a secure ad hoc network and to apply the authentication principle used by SRP, we added several changes and new parameters to SRP. Each node maintains a neighbor table that contains information (parameters used to calculate the secret) about neighboring nodes. In ASRoP protocol, we kept the same format of messages (*Hello, route request, route reply and route error*) used in AODV, and we added two other kinds of messages (*KeyExchange* and *Authentication*).

3.3.1 Format of KEYEXCHANGE packets

These packets (Figure 5) are used during the first step of ASRoP to share the secret key using the Diffie-Hellman algorithm; this secret key is used as a password during the second step. Each KEYEXCHANGE packet includes the following information:

- **IP@_D, @IP_S**: destination and source IP Addresses;
- **Prime**: a very large prime number;
- **Generator**: integer used in calculations;
- **Public Key**: the public key of a node calculated according to the Diffie-Hellman algorithm;
- **Confirmation**: indicates whether the packet is a request or a reply.

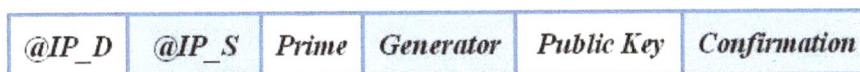

@IP_D	@IP_S	Prime	Generator	Public Key	Confirmation

Figure 5. Exchanges key packets: KEYEXCHANGE

3.3.2 Authentication packet: AUTH

This packet (Figure 6) is used to exchange useful parameters to perform a mutual authentication between two nodes that share the same secret according to the principle of SRP. AUTH packet includes the following information:

- **IP@_D, @IP_S:** destination and source IP addresses;

- **Reply_Dest:** destination IP address of the reply packet route;

- **Type:** authentication packet type (*authentication request, authentication response, request for verification* or *authentication success*);

- **Parameters:** values used in the calculation and verification of authentication.

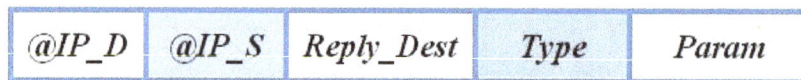

@IP_D	@IP_S	Reply_Dest	Type	Param

Figure 6. Format of AUTH packets

3.4. ASRoP analysis

In the following, we provide a security analysis of the proposed ASRoP by evaluating its robustness in the presence of attacks. The proposed protocol is efficient for negotiating secure connections using a password or a key, while eliminating some security issues typically associated with reusable passwords. ASRoP performs a secure key exchange during the authentication process, allowing security layers (privacy protection and/or integrity) to be activated during the session lifetime. Key servers and reliable certificate infrastructures are not required in our solution, and clients should not store or manage long-term keys. ASRoP offers both security and deployment advantages over existing challenge-response techniques, making it an ideal drop-in replacement where the authentication of a secure password is required [k1].

It is impossible to deduce the key from another legitimate node because it is the only one which knows its value, even if an attacker listens on the communication medium he cannot deduce the key since the latter is never transmitted over the channel (except in the first exchange which we assume as secure).The advantage of the proposed protocol lies in the fact that a node cannot participate in a process of route discovery only if he owns the rights. Hence, unauthenticated nodes cannot send control messages.

If an internal node tries to spoof the destination identity by responding to the source node, its message will inevitably be rejected. Indeed, the authentication process will not be realized. This ensures that the established route between two nodes is always valid and cannot be changed by an attacker.

In this section, we have explained in details our proposed protocol which ensures the routing security in MANETs to guaranty the communication of user's data despite the presence of intruders and attacks related to routing. However, our protocol is still not perfect; if a legitimate node refuses to cooperate (selfishness), the authentication is no longer sufficient and we need more advanced mechanisms such as reputation systems. We can say that the proposed ASRoP protocol is a first line of defense against any external intrusion.

3. SIMULATION AND PERFORMANCE EVALUATION

In the following, we conduct a simulation study using ns-2 to evaluate and compare the performance of our proposed protocol, i.e., ASRoP, with AODV scheme. Two scenarios will be illustrated in order to properly evaluate our protocol: in the first scenario, the velocity is set to 1 m/s which represents the velocity of walking man, whereas in the second one, the velocity is set up to 25 m/s which is dedicated for vehicular case. The performance evaluation involves the metrics which are often used in the evaluation of routing security protocols:

- *End to End Delay*: this parameter represents the average time for transmitting a data packet from source to destination. It introduces all the delays for the route establishment. This metric is very interesting in many applications (e.g. real-time voice traffic) requiring a critical delay.

- *Packet Delivery Ratio*: it is the rate of packets successfully delivered. This metric represents the percentage of packets delivered successfully to their destinations compared to the sum of data packets transmitted in the network.

- *Route Acquisition Delay*: it is the average time for discovering a route between a source and a destination.

- *Average Path Length*: it represents the average number of hops between a source and a destination; often used as a metric for choosing the optimal path (shortest path in terms of number of hops).

- *Dropped packets*: it is the number of data packets lost due to network congestion or packets collision.

- *Routing Load*: this parameter represents the quantity of control packets generated by the protocol to discover and maintain a route.

The simulations are done on *NS-2 version 2.34* in a *Linux* environment. The used models in the simulation are standard and have the following properties:

- **Antenna Model**

We used an omnidirectional antenna that broadcasts to 360° around it. With this model a node can communicate with all its neighbors in any direction, unlike the type of directional antenna that requires that the transmitter antenna is pointing in the direction of the receiving antenna. The radio range is fixed at 250 m, which is a realistic value considered by existing wireless cards.

- **Propagation model**

The propagation model informs us about how the signals will be attenuated according to distance. For example, the *free space model* considers the ideal case where there is only one propagation path between transmitter and receiver and it is in direct view, while the *Two-ray ground model* considers both the direct path and a reflection on the ground.

- **Traffic model**

Usually, a generated traffic in network has to consider several parameters. We set our parameters as follows: the size of a packet is equal to 512 bytes and the sending frequency is 4 packets per second; thus, the flow rate of each source is equal to 4*512*8 bit/sec = 16 kbits/sec. The number of connections is set to 5 to avoid overloading the network. Since the purpose of our simulations is to analyze the properties of the proposed solution, traffic sources generate a constant rate *CBR (constant bit rate)*.

- **Mobility Model**

Varying the characteristics of mobility conveys a significant impact on routing performance. In our simulated network, mobile nodes move according to the RWP model (*Random Waypoint Model*). This model is widely used in research in mobile wireless networks. It provides scenarios where all mobile nodes move randomly during simulation.

4.1. Performance study

We have studied the behavior of the proposed protocol on a large scale by changing the number of nodes. All simulations are repeated 250 times and a confidence interval of 0.95 was considered. Two steps are considered to assess the performance of ASRoP: the first one investigates the impact of network density on ASRoP; while in the second step, we study the performance of ASRoP under various nodes velocity.

4.1.1   Scenario 1: variation of network density

Table 1. Simulation parameters in scenario 1

Parameters	Values
Antenna	OmniAntenna
Number of Nodes	10-100 nodes
Mobility	1 m/s
MAC layer type	IEEE 802.11
Radio propagation model	Two ray ground
Mobility model	Random way point
CBR traffic	4 packets/s
Packets size	512 bits
Pause time	10 sec
Network dimension	1000 * 1000
Transmission range	250 m
Simulation time	250 sec

4.1.1.1 End to end delay and route discovery time

Figure 7 shows an increasing trend in the average time from start to finish depending on the number of nodes. We can observe that the average time of ASRP protocol is significantly higher than that of AODV protocol. This is due to the time required by the ASRP operations used for authentication which makes the route discovery process slower than in AODV. However, this is not dramatic because it does not impact the protocol performance.

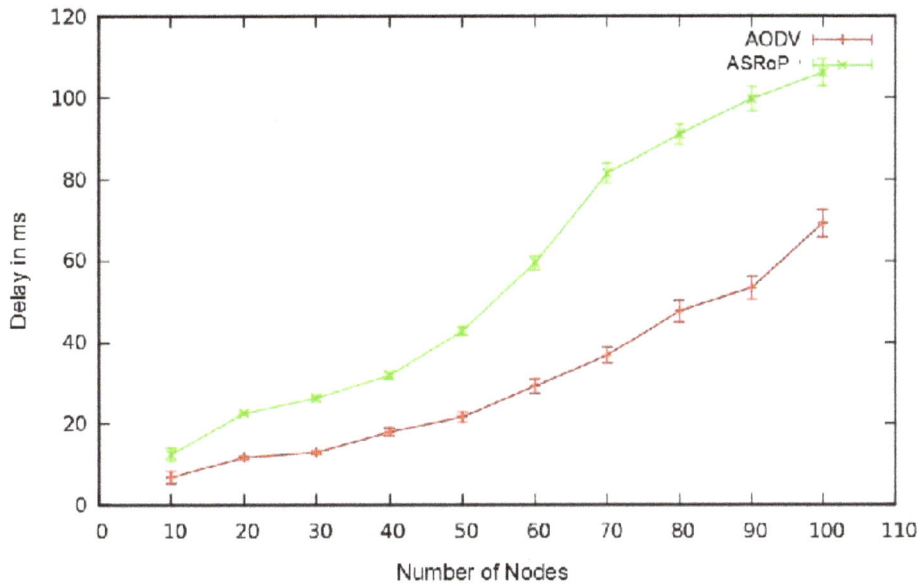

Figure 7. End to end delay

We notice the same trend in Figure 8 which illustrates the evolution of the route discovery time based on the number of nodes. The figure shows that the route discovery in ASRP requires more time than in AODV. In the case of 50 nodes, the route discovery time is 0.49 ms using AODV and 1.26 ms using ASRP. In the case of 100 nodes, the route discovery time is 3.19 ms using AODV and 4.56 ms using ASRP. However, this time is in milliseconds, we can say that ASRP allows route discovery in a reasonable time.

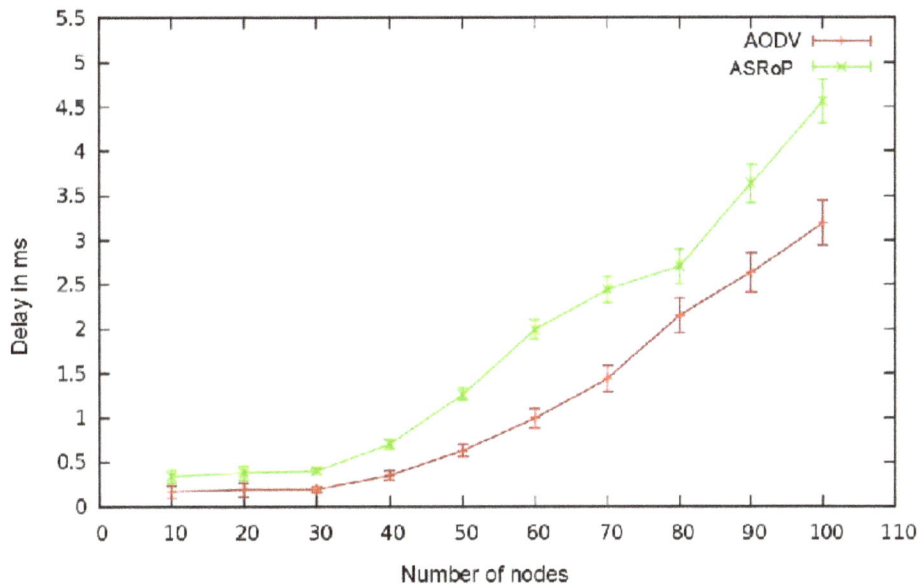

Figure 8. Route discovery time

4.1.1.2 Packets loss

By analyzing the Figures 9 and 10, we can say that the considered metrics (successful delivery rate and packets loss rate) do not make differences between the ASRP and AODV protocols. The two curves, in both figures, have the same trend. The rate of successfully delivered packets varies between 0.97 and 0.99. This high rate implies that the delivery of data packets is achieved successfully, thus the performance was not degraded even with the presence of security extensions.

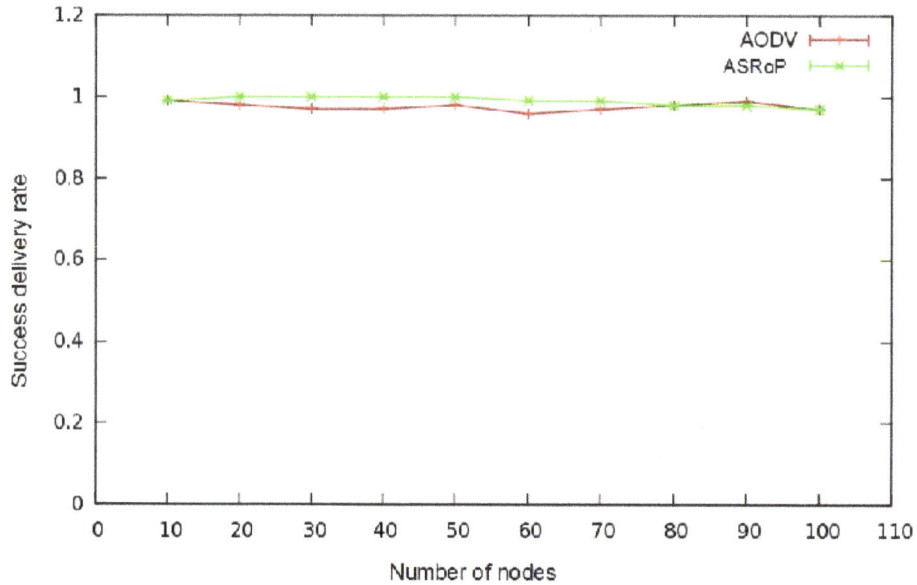

Figure 9. Successful delivery rate

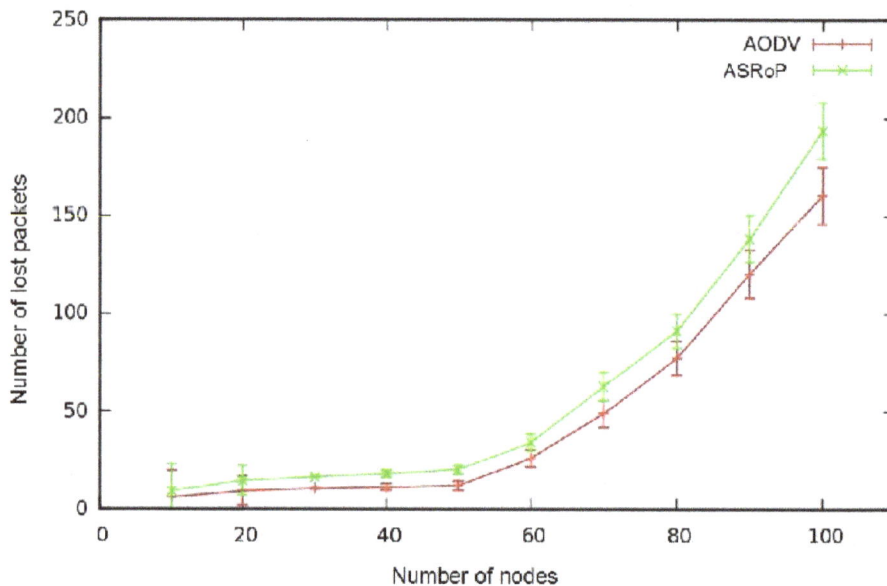

Figure 10. Packet Loss rate

We remark also that small performance degradation occurs by increasing the number of nodes: the scaling leads to an increase of exchanged messages, which causes an increase in the number of lost packets. This is often due to collisions between packets.

4.1.1.3 Number of hops by route and routing overhead

Traditionally, the shortest path in terms of number of hops is considered as the best path to route data. The AODV protocol uses this metric to choose its routes. We assumed that the first path obtained is the best one. This means that updating the routing table is due to the arrival of the first response (the fastest one). In Figure 11, we notice that the curves in the two protocols are very close. The obtained average number of hops proving that the length of routes established by ASRP is not necessarily longer than that of AODV.

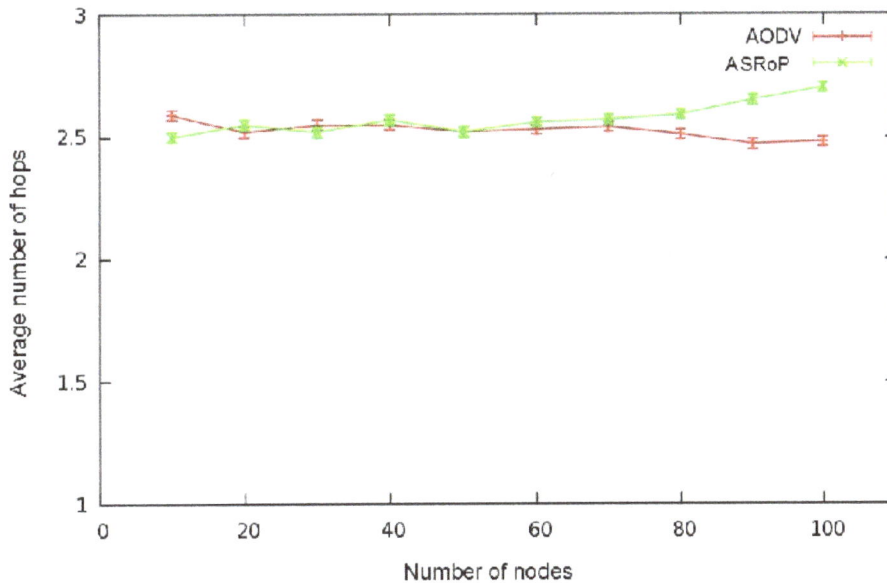

Figure 11. Number of hops/route

Figure 12 shows the number of control packets in function of the number of nodes. This parameter let us observe the cost of a protocol in terms of resource consumption and enable to understand the resistance of the protocol in case of congestion. In both protocols the variations of the number of nodes has a direct influence on the number of exchanged packets; this is due to the diffusion of various control messages. However, the number of exchanged control packets is greater in ASRP that in AODV. This is mainly due to the new extension realized in ASRP, which requires in the first step to share the keys and in the second one to run the authentication process.

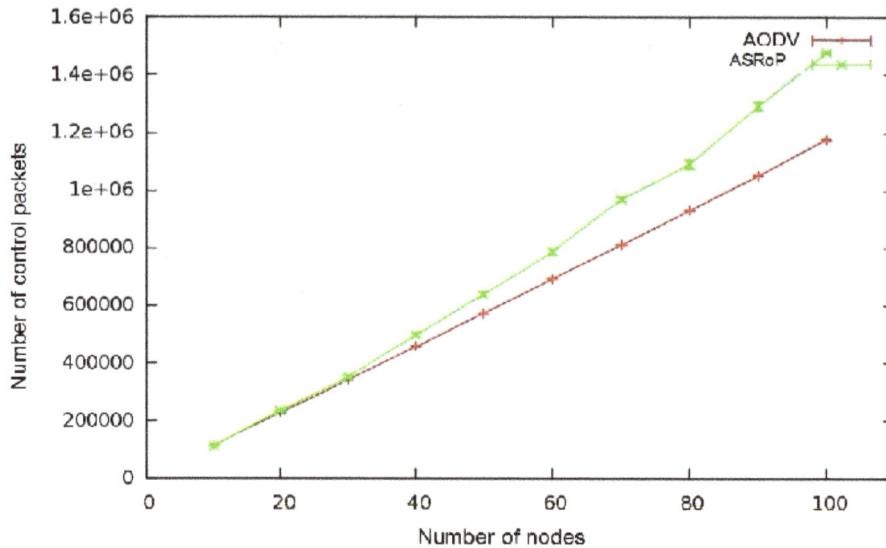

Figure 12. Number of control packets

4.1.2   Scenario 2: variation of nodes velocity

In this section, we study the impact of nodes velocity on the ASRoP. The simulation parameters are listed in the table 2.

Table 2. Simulation parameters in scenario 2

Parameters	Values
Antenna	OmniAntenna
Number of Nodes	25 nodes
Mobility	1-25 m/s
MAC layer type	IEEE 802.11
Radio propagation model	Two ray ground
Mobility model	Random way point
CBR traffic	4 packets/s
Packets size	512 bits
Pause time	10 sec
Network dimension	1000 * 1000
Transmission range	250 m
Simulation time	250 sec

4.1.2.1 End to end delay and route discovery time

Figure 13 shows an increasing trend in the average end-to-end delay depending on the mobility of nodes. We can observe that the average delay in ASRoP is significantly higher than that in AODV. This is due to the time required by the cryptographic procedures and messages exchanged making the route discovery process slow.

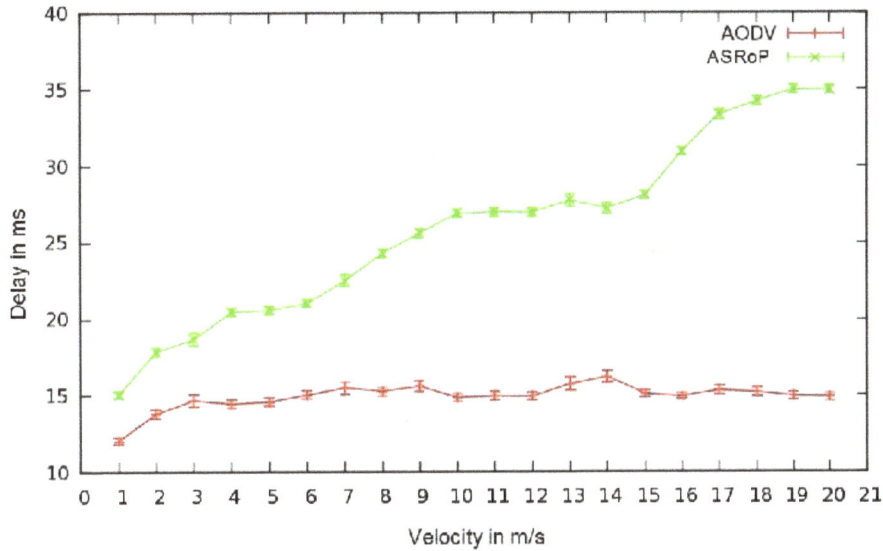

Figure 13. End to end delay

We observe the same behavior in Figure 14 which represents the evolution of the route discovery time in function of nodes mobility. In this figure, we can observe that the route acquisition in ASRoP requires more time than in AODV. This figure illustrates the strong impact of mobility on ASRoP protocol: having high mobility leads to more control messages, since there is less connectivity (nodes can get away from each others). However, this time as it is in milliseconds, we can say that ASRoP allows getting a secure route in a reasonable time.

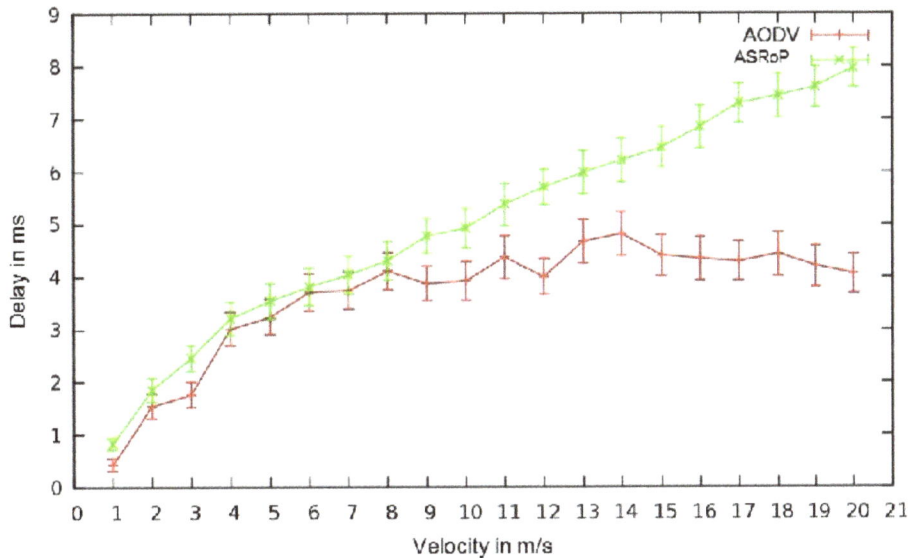

Figure 14. Route discovery time

4.1.2.2 Packets loss

By observing the two graphs in 15 and 16, we can say that these two metrics do not illustrate a difference between the two protocols; the two curves have almost the same trend. The rate of successfully delivered packets varies between 0.96 and 0.99. This high rate implies that the delivery of data packets is very successful, so the performance was not degraded even with the presence of security extensions.

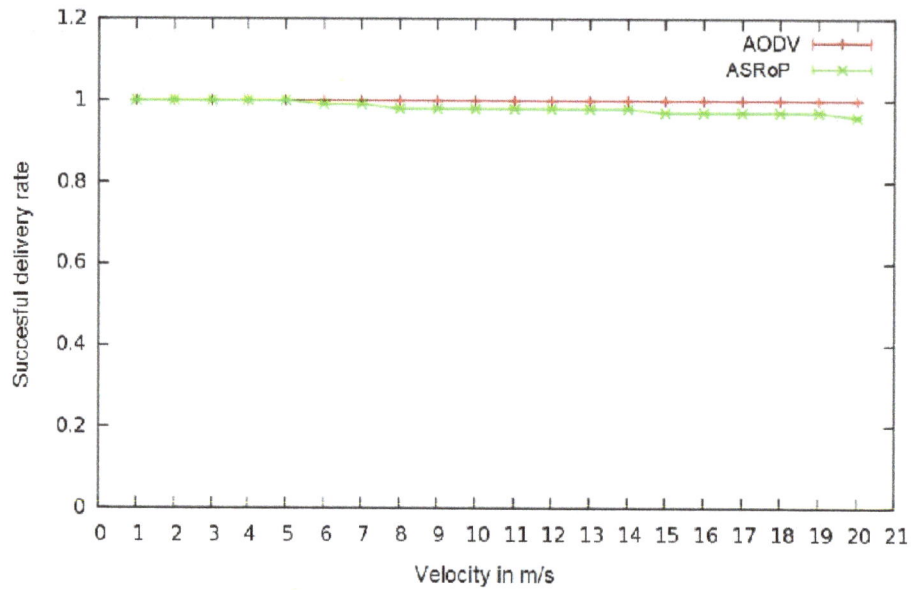

Figure 15. Successful delivery rate

We observe that small performance degradation occurs when increasing the mobility of nodes: the node's mobility leads to augment the frequency of broken links, thereby to increase the number of lost packets. This loss is illustrated in Figure 16 but it is controlled by the routing algorithm that tries to adapt to mobility through a local repair of the broken paths.

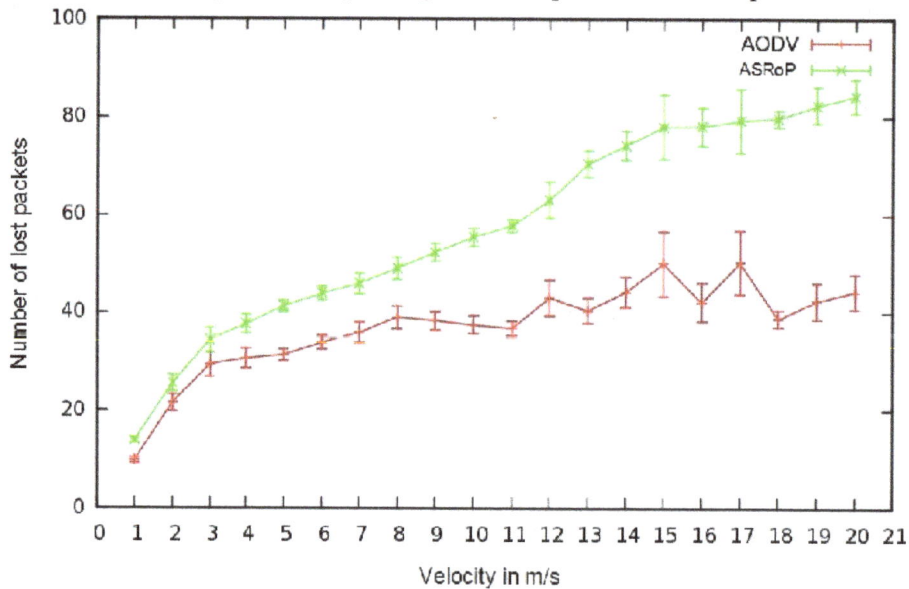

Figure 16. Packets loss rate

4.1.2.3 Number of hops by route and routing overhead

In Figure 17, we notice that the curves of the two protocols are very close. The obtained average number of hops proves that the length of routes established by ASRoP is not necessarily longer than that of AODV.

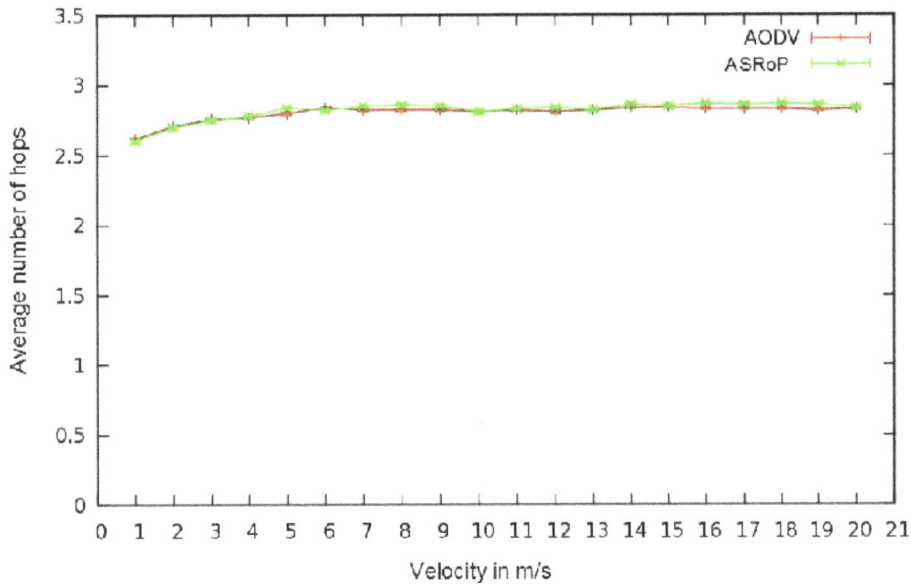

Figure 17. Number of hops/route

In Figure 18, in both protocols the mobility has a direct influence on the number of exchanged packets. This is mainly due to the diffusion of various control messages; however, their number is greater in ASRoP than in AODV. This is due to the new extension which requires a first step to share the keys and a second to run the authentication process. The number of control messages decreases when the mobility increases, as we can observe by comparing the two Figures 18 and 12.

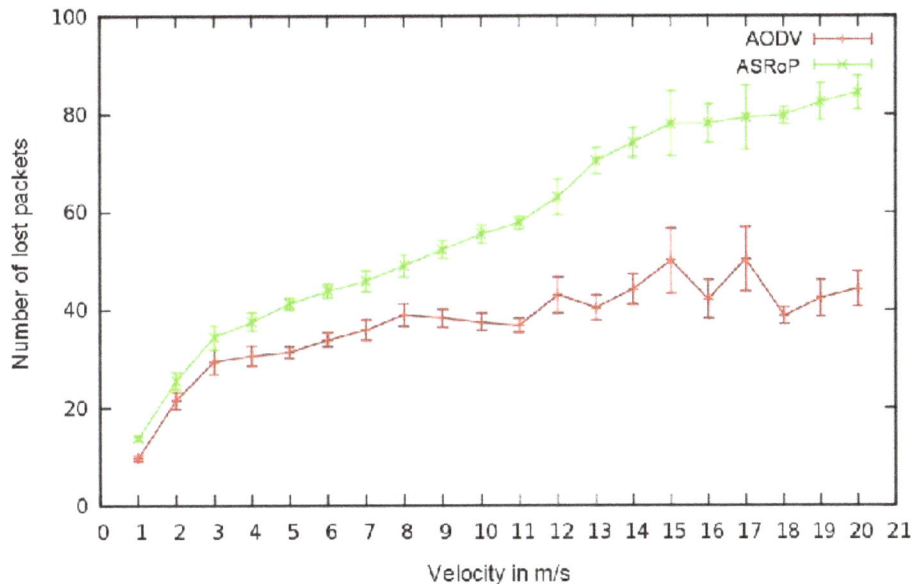

Figure 18. Number of control packets

Through these results, we can clearly see the impact of mobility and variation of the number of nodes on the evaluated metrics. The simulations have shown that compared to AODV, our security solution consumes slightly more resources and that the delays are longer is ASRoP than in AODV. Indeed, in each hop, the exchange of messages, which ensures authentication increases the security cost of ASRoP.

5. CONCLUSIONS AND PERSPECTIVES

In this paper, we have analyzed some security attacks a MANET may face, described some ad hoc routing security solutions, and presented a new secure routing protocol named ASRoP. The idea that we consider most appropriate to ensure secure communications between entities in an ad hoc network is the establishment of a cryptographic mechanism to ensure authentication of messages. The proposed ASRoP is based on Diffie-Hellman algorithm and SRP authentication protocol that we have exploited to negotiate a secure session using a user password, while eliminating the security concerns often associated with reusable passwords. The performance evaluations have proved that ASRoP offers good results; specifically in terms of the rate of successful delivery packets and end-to-end delay of packets transmission. The ASRoP results are almost the same as for AODV protocol, with the advantage of securing communications in ASRoP. However, a comparison with other secure AODV protocols (e.g. [21] [9]) would be very useful in order to demonstrate efficiently the robustness of ASRoP. In a future work, we plan to add a reputation mechanism for ASRoP and evaluate the proposed ASRoP with other parameters. In addition, simulating realistic attack scenarios would be very interesting to test the robustness of ASRoP in practical cases.

REFERENCES

[1]   Charles E. Perkins, Elizabeth M. Belding-Royer, and Samir R. Das. Ad hoc on-demand distance vector (aodv) routing. RFC Experimental 3561,Internet Engineering Task Force, July 2003.

[2]   David B. Johnson and David A. Maltz. Dynamic source routing in ad hoc wireless networks. In Thomasz Imielinski and Hank Korth, editors, Mobile Computing, volume 353, chapter 5, pages 153_181. Kluwer Academic Publishers, 1996.

[3]   Ha Duyen Trung, Watit Benjapolakul, and Phan Minh Duc. Performance evaluation and comparison of di_erent ad hoc routing protocols. Comput. Commun., 30:2478_2496, September 2007.

[4]   L. Abusalah, A. Khokhar, and M. Guizani. A survey of secure mobile ad hoc routing protocols. Communications Surveys Tutorials, IEEE, 10(4):78_93, 2008.

[5]   P. Sakarindr and N. Ansari. Security services in group communications over wireless infrastructure, mobile ad hoc, and wireless sensor networks. Wireless Communications, IEEE, 14(5):8 _20, october 2007.

[6]   Dongbin Wang, Mingzeng Hu, and Hui Zhi. A survey of secure routing in ad hoc networks. In Proceedings of the 2008 The Ninth International Conference on Web-Age Information Management, WAIM '08, pages 482_486, Washington, DC, USA, 2008. IEEE Computer Society.

[7]   G Xu, C Borcea, and L Iftode. A policy enforcing mechanism for trusted ad hoc networks. Dependable and Secure Computing, IEEE Transactions on, 2010.

[8]   K. Driscoll, B. Hall, M. Paulitsch, P. Zumsteg, and H. Sivencrona. The real byzantine generals. In Digital Avionics Systems Conference, 2004. DASC 04. The 23rd, volume 2, pages 6.D.4 _ 61_11 Vol.2, 2004.

[9]   Ming Yu, Mengchu Zhou, and Wei Su. A secure routing protocol against byzantine attacks for manets in adversarial environments. Vehicular Technology, IEEE Transactions on, 58(1):449_460, january 2009.

[10]   Hakima Chaouchi and Maryline Laurent-Maknavicius. La sécurité dans les réseaux sans fil et mobiles, Tome 2, Technologies du marché. April 2007.

[11]   M. Krasnovsky and V. Wieser. A performance of wireless ad-hoc network routing protocol. In Radioelektronika, 2007. 17th International Conference, pages 1 _3, 2007.

[12]   Hu. Yih-Chun and A. Perrig. A survey of secure wireless ad hoc routing. Security Privacy, IEEE, 2(3):28 _39, 2004.

[13]   Sridhar Radhakrishnan, Gopal Racherla, and David Furuno. Mobile ad hoc networks: principles and practices, pages 381_405. CRC Press, Inc., Boca Raton, FL, USA, 2003.

[14]   L. Tamilselvan and V. Sankaranarayanan. Solution to prevent rushing attack in wireless mobile ad hoc networks. In Ad Hoc and Ubiquitous Computing, 2006. ISAUHC '06. International Symposium on, pages 42 _47, 2006.

[15]   Ashish Raniwala, Ashish Raniwala, and Ashish Raniwala. Architecture and protocols for a high-performance, secure ieee 802.11-based wireless mesh network, 2009.

[16]   Youngho Park, Won-Young Lee, and Kyung-Hyune Rhee. Authenticated on-demand ad hoc routing protocol without pre-shared key distribution. In Proceedings of the 2007 ECSIS Symposium on Bio-inspired, Learning, and Intelligent Systems for Security, pages 41_46, Washington, DC, USA, 2007. IEEE Computer Society.

[17]   K. Sanzgiri, D. LaFlamme, B. Dahill, B.N. Levine, C. Shields, and E.M. Belding-Royer. Authenticated routing for ad hoc networks. Selected Areas in Communications, IEEE Journal on, 23(3):598 _ 610, 2005.

[18]   P. Papadimitratos and Z.J. Haas. Secure link state routing for mobile ad hoc networks. In Applications and the Internet Workshops, 2003. Proceedings. 2003 Symposium on, pages 379 _ 383, 2003.

[19]   Yih-Chun Hu, D.B. Johnson, and A. Perrig. Sead: secure e_cient distance vector routing for mobile wireless ad hoc networks. In Mobile Computing Systems and Applications, 2002. Proceedings Fourth IEEE Workshop on, 2002.

[20]   ThomasWu. The secure remote password protocol. In In Proceedings of the 1998 Internet Society Network and Distributed System Security Symposium, pages 97_111, 1998.

[21]   D. Cerri and A. Ghioni. Securing aodv: the a-saodv secure routing prototype. Communications Magazine, IEEE, 46(2):120 _ 125, february 2008.

11

A Multi-Hop Weighted Clustering of Homogeneous Manets Using Combined Closeness Index

T.N. Janakiraman[1] and A. Senthil Thilak[2]

[1, 2]Department of Mathematics, National Institute of Technology, Tiruchirapalli-620015, Tamil Nadu, India
[1]janaki@nitt.edu, tnjraman2000@yahoo.com
[2]asthilak23@gmail.com

ABSTRACT

In this paper, a new multi-hop weighted clustering procedure is proposed for homogeneous Mobile Ad hoc networks. The algorithm generates double star embedded non-overlapping cluster structures, where each cluster is managed by a leader node and a substitute for the leader node (in case of failure of leader node). The weight of a node is a linear combination of six different graph theoretic parameters which deal with the communication capability of a node both in terms of quality and quantity, the relative closeness relationship between network nodes and the maximum and average distance traversed by a node for effective communication. This paper deals with the design and analysis of the algorithm and some of the graph theoretic/structural properties of the clusters obtained are also discussed.

KEYWORDS

Homogeneous, Mobile Ad hoc networks, Double star, Leader node, Relative closeness relationship

1. Ad Hoc Networks – A Brief Review

An ad hoc wireless network is a collection of two or more devices (also termed as nodes) equipped with wireless communications and networking capability. Such devices/nodes can communicate either directly or through intermediate nodes depending on the availability of the nodes within or outside the radio range. An ad hoc network is self-organizing and adaptive, i.e, the already formed network can be de-formed on-the-fly without the need for any central administration. The nodes in an ad hoc network must be capable of identifying the connectivity with the neighbouring nodes, so as to allow communication and sharing of information and services. The nodes must perform routing and packet-forwarding functions. The topology changes continuously as the devices are not tied down to specific locations over time. Hence, the most important and challenging issues in a mobile ad hoc network are the mobile nature of the devices, scalability and constraints on resources such as limited bandwidth, limited and varying battery power, etc. Depending on the nature of devices, the uniformity in transmission range and network architecture, the network can either be homogeneous or heterogeneous. The network considered in this paper is a homogeneous where each node is assumed to have uniform transmission range.

2. Significance of Clustering

A *cluster* is a subset of nodes of a network. *Clustering* is the process of partitioning a network into clusters and it is a way of making ad hoc networks more scalable. Scalability refers to the network's capability to facilitate efficient communication even in the presence of large number of network nodes. Cluster-based structures promote more efficient usage of resources in controlling large dynamic networks. With cluster-based control structures, the physical network is transformed into a virtual network of interconnected node clusters. Clustering can be done for different purposes, such as, clustering for transmission management, clustering for backbone management, clustering for routing efficiency etc, [1]. Each cluster has one or more controllers, such as leader nodes(also called as Masters or cluster-heads), Proxy nodes, Super-masters, gateways, etc. [2], acting on its behalf to make control decisions for cluster members and in some cases, to construct and distribute representations of cluster state for use outside of the cluster. The algorithm proposed in this paper is developed with the objective of facilitating routing functions by providing a hierarchical network organization and efficient sharing of resources and information.

In general, the process of clustering involves two phases, namely, cluster formation and cluster maintenance. Initially, the nodes are group together based on some principle to form the clusters. Then, as the nodes continuously move in different directions with different speeds, the existing links between the nodes also get changed and hence, the initially formed cluster structure cannot be retained for a longer period. So, it is necessary to go for the next phase, namely, cluster maintenance phase. Maintenance includes the procedure for modifying the cluster structure based on the movement of a cluster member outside an existing cluster boundary, battery drainage of cluster-heads, link failure, new link establishments, addition of a new node, node failure and so on.

3. Prior Work

Several procedures are proposed and adopted for clustering of mobile ad hoc networks. Weight based clustering algorithms [4-7], Zone based clustering algorithms [8, 9], Dominating set based clustering [9, 10, 11] etc., are to name a few. In these clustering procedures, the clusters are formed based on different criteria and the algorithms are classified accordingly.

Based on whether a special node with specific features is required or not, the algorithms can be classified as cluster-head based and non-cluster-head based algorithms [11, 12]. Based on the hop distance between different pair of nodes in a cluster, they are classified as 1-hop clustering and multi-hop clustering procedures [11, 12]. Similarly, there exists a classification based on the objective of clustering, such as Dominating set based clustering, low maintenance clustering, mobility-aware clustering, energy efficient clustering, load-balancing clustering, combined-metrics based/weight based clustering [11]. This paper gives another different approach for clustering of such networks. As discussed in [13], the proposed algorithm is a weight based cum multi-hop clustering algorithm and is also an extension of 3hBAC [14] and LCC [15] clustering procedures. Hence, we give an overview of some of the algorithms coming under the two categories.

LID Heuristic. This is a cluster-head based, 1-hop, weight based clustering algorithm proposed by Baker and Ephremides [16, 17]. This chooses the nodes with lowest id among their neighbors as cluster-heads and their neighbors as cluster members. However, as it is biased to choose nodes with smaller ids as cluster-heads, such nodes suffer from battery drainage resulting in shorter life span. Also, because of having lowest id, a highly mobile node may be elected as a cluster-head, disturbing the stability of the network.

HD Heuristic. The highest degree (HD) heuristic proposed by Gerla et al. [18, 19], is again a cluster-head based, weight based, 1-hop clustering algorithm. This is similar to LID, except that node degree is used instead of node id. Node Ids are used to break ties in election. This algorithm doesn't restrict the number of nodes ideally handled by a cluster-head, leading to shorter life span. Also, this requires frequent re-clustering as the nodes are always under mobility.

Least cluster change clustering (LCC) [15]. This is an enhancement of LID and HD heuristics. To avoid frequent re-clustering occurring in LID & HD, the procedure is divided into two phases as in the proposed algorithm. The initial cluster formation is done based on lowest ids as in LID. Re-clustering is invoked only at instants where any two cluster-heads become adjacent or when a cluster member moves out of the reach of its cluster-head. Thus, LCC significantly improved stability but the second case for re-clustering shows that even the movement of a single node (a frequent happening in mobile networks) outside its cluster boundary will cause re-clustering.

3-hop between adjacent clusterheads (3hBAC). The 3hBAC clustering algorithm [14] is a 1-hop clustering algorithm which generates non-overlapping clusters. It assigns a new status, by name, cluster-guest for the network nodes apart from cluster-head and cluster member. Initially, the algorithm starts from the neighborhood of a node having lowest id. Then, the node possessing highest degree in the closed neighbor set of the above lowest id is elected as the initial cluster-head and its 1-hop neighbors are assigned the status of cluster members. After this, the subsequent cluster formation process runs parallely and election process is similar to HD heuristic. The cluster-guests are used to reduce the frequency of re-clustering in the maintenance phase.

Weight-based clustering algorithms. Several weight-based clustering algorithms are available in the literature [4-7], [14], [20], [21]. All these work similar to the above discussed 1-hop algorithms, except that each node is initially assigned a weight and the cluster-heads are elected based on these weights. The definition of node weight in each algorithm varies. Some are distributed algorithms [4], [20], [6] and some are non-distributed [5], [7], [14]. Each has its own merits and demerits.

DSECA [13]. The DSCEA is also a weight-based clustering which generates double star embedded non-overlapping structures, where the weight of each node is a linear combination of six parameters, namely, degree, node closeness index, mean hop distance, mean Euclidean distance and neighbour strength value. The algorithm proposed in this paper is a modified version of DSECA.

4. MODELLING ASSUMPTIONS

It is assumed that the network to be clustered is deployed by distributing the mobile nodes randomly in different positions on a terrain of size KxK.. Each node is assumed to have a uniform transmission range and the network under consideration is assumed to be homogeneous, unless otherwise specified. Those nodes within the transmission range of a particular node are identified as the 1-hop neighbors of that node. Each node identifies its 1-hop neighbors by transmitting Hello messages. The nodes are allowed to move randomly in different directions with varying velocity in the range [0, V_{max}]. To keep track of the changes in node positions due to mobility, the nodes send and receive Hello messages periodically at a predefined broadcast interval BI.

Each node computes its own weight and broadcasts a Weight_info() message containing its id, weight. Upon successful transmission and reception of Weight_info() messages by the entire set of nodes, each node maintains a weight table containing the weight information about all the other nodes in the network. Further, each node in the network has knowledge about the hop and Euclidean distance between itself and all the other nodes in the network. With these basic assumptions and information, the network nodes execute our proposed clustering procedure.

5. GRAPH PRELIMINARIES

A *graph G* is defined as an ordered pair *(V, E)*, where *V* is a non-empty set of vertices/nodes and *E* denotes the set of edges/links between different pairs of nodes in *V*. Communication networks can in general be modeled using graphs. If any two nodes are within the transmission range of each other, then both can communicate with each other and are joined by a bidirectional link. The set of all nodes in the network is taken as the vertex (or node) set *V* of *G* and any two nodes are made adjacent (i.e., joined by a link) in *G*, if the corresponding two nodes can communicate with each other and the graph so obtained is called the *underlying graph or network graph or network topology*. Hence, the problem of *"Network Clustering"* can be viewed as a problem of *"Graph Partitioning"*. Since each node is assumed to have uniform transmission range the underlying graph will always be an undirected graph.

If u and v are any two nodes in the network graph, then *d(u, v)* denotes the least number of hops to move from u to v and vice versa and is referred to as the *Hop-distance* between u and v and *ed(u, v)* denotes the Euclidean distance between u and v. Thus, in a homogeneous network, for a given transmission range r, two nodes u and v can communicate with each other only if they are at Euclidean distance less than or equal to r i.e., *ed(u, v) ≤ r*. Graph theoretically, two nodes u and v are joined by a link *e = (u, v)* or made adjacent in the network graph if their Euclidean distance is less than or equal to r, i.e., *ed(u, v) ≤ r*, else they are non-adjacent. The nodes u and v are called the *end nodes* of the link *e = (u, v)*. For a given node u, the neighbor set of u, denoted by *N(u)*, is the set of those nodes which are within the transmission range of u, i.e, the set of those nodes which are 1-hop away from u and the cardinality of the set *N(u)* is defined as the *degree* of u and is denoted by *deg(u)*. The hop-distance between u and its farthest node in G is called the *eccentricity of u* in G and is denoted by *ecc(u)*, i.e., $ecc(u) = \max_{v \in V(G)} \{d(u,v)\}$. The average of the Hop-distances between u and each of the other nodes is defined to be *the mean-hop-distance of u* and is denoted by *MHD(u)* i.e., $MHD(u) = \frac{1}{|V|}\left[\sum_{v \in V(G)} d(u,v)\right]$. The average of the Euclidean distances between u and each of the other node is defined to be *the mean Euclidean distance of u* and is denoted by *MED(u)* i.e., $MED(u) = \frac{1}{|V|}\left[\sum_{v \in V(G)} ed(u,v)\right]$. The minimum and maximum eccentricities of *G* are defined respectively as *radius r(G)* and *diameter d(G)* of G .

A subset *S* of vertices of a graph *G* is said to be a *dominating set of G* if each vertex in *V-S* is adjacent to atleast one vertex in *S*. An *edge e = (u, v)* is said to *dominate an edge f* if either *f = (u, x)* or *f = (v, x)*, where *x ∈ V*. In other words, the edge *e = (u, v)* dominates an edge *f*, if *f* has atleast one of the vertices u or v as one of its end vertices [22]. An edge subset *E′⊆ E* is an *efficient edge dominating set* for *G* if each edge in *E* is dominated by exactly one edge in *E′* [22].

The graph *G* which is rooted at a vertex say v, having n nodes $v_1, v_2, ... v_n$, adjacent to v as shown in Figure 1, is called as the *star graph* and is denoted by $K_{1, n}$.

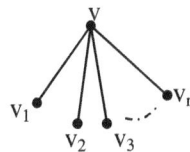

Figure 1. Star Graph

The graph obtained by joining the root vertices of the stars $K_{1, n}$ and $K_{1, m}$ by means of an edge as shown in Figure 2, is referred to as a *double star* or *(n, m)-bi-star.*

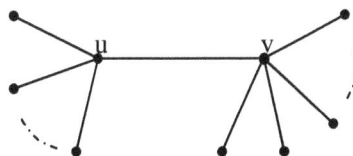

Figure 2. Double Star/(n, m)-bi-star

6. DEFINITION OF WEIGHT PARAMETERS

With the support of the idea generalized in [3], i.e., any meaningful parameter can be used as the weight to best exploit the network properties, here we use six different graph theoretic parameters for computing the weight of each node.

6.1. Closer-Hop set

Given a pair of nodes u and v in a graph G, the *closer-hop set of u relative to v,* is defined as the set of those nodes which are at a shorter hop distance with u compared to v and is denoted by $CHS(u|v)$, i.e., $CHS(u|v) = \{w \in V(G) : d(u, w) < d(v, w)\}$ and $c_h(u|v)$ is the cardinality of the $CHS(u|v)$. It is to be noted that $c_h(u|v)$ need not be equal to $c_h(v|u)$. In fact, $c_h(v|u) = N - c_h(u|v)$, where N denotes the total number of nodes in the network.

6.2. Closer-Euclidean set

Given a pair of nodes u and v in G, the *closer-euclidean set of u relative to v,* is defined as the set of those nodes which are at a shorter euclidean distance with u compared to v and is denoted by $CES(u|v)$, i.e., $CES(u|v) = \{w \text{ in } V(G) : ed(u, w) < ed(v, w)\}$ and $c_{ed}(u|v)$ is the cardinality of the $CES(u|v)$.

6.3. Hop-Closeness Index

Given two nodes u and v in a graph G, if $f_h(u, v) = c_h(u|v) - c_h(v|u)$, then the *hop-closeness index of u* denoted by $g_h(u)$, is defined as $g_h(u) = \sum_{v \in V(G)-u} f_h(u,v)$. For example, consider the graph in Figure 3. The Hop-closeness index of the node 1 is calculated as follows.

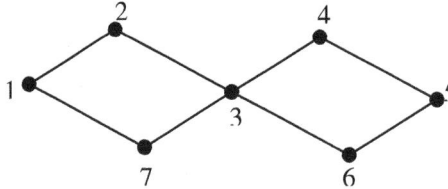

Figure 3. Graph for computing closeness index

$$g_h(1) = \sum_{i \in V(G)-1} f_h(1,i)$$
$$= f_h(1, 2) + f_h(1, 3) + f_h(1, 4) + f_h(1, 5) + f_h(1, 6) + f_h(1, 7)$$
$$= (-3) + (-1) + (0) + (-1) + (-3) + (-3)$$
$$= (-11)$$

6.4. Euclidean-Closeness Index

Given two nodes u and v in a graph G, if $f_{ed}(u, v) = c_{ed}(u|v) - c_{ed}(v|u)$, then the *euclidean-closeness index of u* denoted by $g_{ed}(u)$, is defined as $g_{ed}(u) = \sum_{v \in V(G)-u} f_{ed}(u,v)$. By knowing the (x, y) positions of each node in the network, the Euclidean-closeness index of each node can be computed in a similar fashion given in section 6.3.

If $g_h(u)$ (or $g_{ed}(u)$) is positive, then it indicates the positive relative closeness relationship, in the sense that, if for a node u, $g_h(u)$ (or $g_{ed}(u)$) is positive maximum, then it is more closer to all the nodes in the network compared to that of the other nodes. If $g_h(u)$ (or $g_{ed}(u)$) is negative maximum, it indicates that the node u, is highly deviated from all the other nodes in the network compared to that of the others. It gives a measure of the negative relative closeness relationship.

6.5. Combined-closeness index

The Combined-closeness index of a node u, denoted by $CCI(u)$ is defined to be the average of $g_h(u)$ and $g_{ed}(u)$. i.e, $CCI(u) = (g_h(u) + g_{ed}(u))/2$.

6.6. Categorization of neighbours of a node [13]

Depending on the Euclidean distance between the nodes, their signal strength varies. For a given node u (transmitting node), the nodes which are closer to u will receive stronger signals and those nodes which are far apart from u will get weaker signals. Based on this notion, the neighbors of a transmitting node are classified as follows:
  i.   Strong neighbours
  ii. Medium neighbours
  iii. Weak neighbours
Strong neighbour: A node v is said to be a strong neighbour of a node u, if the Euclidean distance between u and v is less than or equal to r. i.e., $0 \leq ed(u, v) \leq r/2$.
Medium neighbour: A node v is said to be a medium neighbour of a node u, if $r/2 \leq ed(u, v) \leq 3r/4$.
Weak neighbour: A node v is said to be a weak neighbour of a node u, if $3r/4 \leq ed(u, v) \leq r$.

6.7. Neighbour Strength value

For any node u in the network, the neighbour strength value denoted by $NS(u)$ is defined to be $NS(u) = (m_1 + m_2/2 + m_3/4)K$, where K is any constant (a fixed threshold value) and m_1, m_2, m_3 denote respectively the number of strong, medium and weak neighbours of u. As explained in [13], for a node u with greater connectivity, its greater value is due to the contribution of all strong, weak and medium neighbours of u. But, if there exists another node v such that $deg(u) > deg(v)$ and $m_1(u) < m_1(v)$, $m_2(u) > m_2(v)$, $m_3(u) >>> m_3(v)$, then it is obvious that node u will be chosen because of having greater connectivity value. But, all its weak neighbours have greater tendency to move away from u. This affects the stability of u and hence the corresponding cluster, if u is chosen as a master/proxy. Hence, we use the parameter $NS(u)$ to determine the quality of the neighbours of a node and hence the quality of the links.

6.8. Node Weight

Since for real time applications, it is better to consider Euclidean distances rather than hop distances in some cases, and the hop distance cannot be ignored completely, in the proposed algorithm, instead of the node closeness index value used in the calculation of node weight in [13], we use the combined-closeness index value, by considering both Euclidean and hop distances. Thus, for any node u in the network, the weight of u, denoted by $W(u)$ is defined as follows.

$$W(u) = \alpha_1 \deg(u) + \alpha_2 CCI(u) + \alpha_3 \left(\frac{1}{ecc(u)}\right) + \alpha_4 \left(\frac{1}{MHD(u)}\right) + \alpha_5 \left(\frac{1}{MED(u)}\right) + \alpha_6 NS(u) \quad ---- (1)$$

Here, the constants $\alpha_1, \alpha_2, \alpha_3, \alpha_4, \alpha_5$ and α_6 are the weighing factors of the parameters under consideration and these may be chosen according to the application requirements. In the proposed algorithm, in order to give equal weightage to all the factors considered, we choose all the weighing factors as $(1/n)$, where n = number of parameters considered. Here, n = 6.

7. STATUS OF THE NODES IN A NETWORK GRAPH

In the proposed algorithm, each node in the network is assigned one of the following status:

* **Master** – A node which is responsible for coordinating network activities and also responsible for inter and intra cluster communication
* **Proxy** – A node adjacent to a master node which plays the role of a master in case of any failure of the master.
* **Slaves** – Neighbors of Master nodes and/or Proxy nodes
* **Type I Hidden Master** – A neighbor node of a Proxy having greater weight than proxy.
* **Type II Hidden Master** – A node with greater weight and eligible for Master/Proxy selection, but not included in cluster formation because of not satisfying distance property and also not adjacent to any Proxy node.
* A node which is neither a slave nor a Master/Proxy.

It is to be noted that a node which was a type II hidden master at some instant may become a type I hidden master at a later instant.

8. BASIS OF OUR ALGORITHM

In all cluster-head based algorithms, a special node called a cluster-head plays the key role in communication and controlling operations. These cluster-heads are chosen based on different criteria like mobility, battery power, connectivity and so on. Though, a special care is taken in these algorithms to ensure that the cluster-heads are less dynamic, the excessive battery drainage of a cluster-head or the movement of a cluster-head away from its cluster members require scattering of the nodes in the cluster structure and re-affiliation of all the nodes in that cluster.

To overcome this problem, in the proposed algorithm, in addition to the cluster-heads (referred to as Masters in our algorithm), we choose another node called Proxy, to act as a substitute for the cluster-head/master, when the master gives up its role and also to share the load of a cluster-head. In our algorithm, each node is assigned a weight based on different criteria. The weight of a node is a linear combination of six different parameters as in (1). The algorithm concentrates on maximum weighted node and the weight is maximum if the parameters deg(u), g(u), NS(u) are maximum and ecc(u), MHD(u) and MED(u) are all minimum. The following characteristics are considered while choosing the parameters.

1. The factor deg(u) denotes the number of nodes that can interact with u or linked to u, which is otherwise stated as the connectivity of the node. By choosing a node u with deg(u) to be maximum, we are trying to choose a node having higher connectivity. This will minimize the number of clusters generated.
2. The metric, neighbour strength value, denoted by NS(u) gives the quality of the links existing between a node and its neighbors. By choosing deg(u) and NS(u) to be simultaneously maximum, we give preference to a node having good quantity and quality of neighbours/links.
3. The parameter CCI(u) gives a measure of the relative closeness relationship between u and the other nodes in the network, both in terms of hop and Euclidean distances. By choosing CCI(u) to be maximum, we are concentrating on the node having greater affinity towards the network.
4. By choosing a node with minimum ecc(u), we concentrate on a node which is capable of communicating with all the other nodes in least number of hops compared to others.
5. By selecting a node with minimum MHD and MED, we choose a node for which the average time taken to successfully transmit the messages (measured both in terms of number of hops and Euclidean distance) among/to the nodes in the network is much lesser.

9. OBJECTIVES OF THE ALGORITHM

The algorithm discussed in this paper is designed with the following objectives.

1. The network nodes are partitioned into different groups of various sizes to form a hierarchical organization of the network.
2. The cluster formation and maintenance overheads should be minimized.
3. The clusters generated must be stable as long as possible.
4. The leader nodes should not be overloaded. Here, it is distributed between the master and proxy nodes.
5. Re-affiliations should be minimized.
6. Re-clustering should be avoided as much as possible. At times of necessity, re-affiliations are allowed instead of re-clustering to reduce the cost of cluster maintenance.
7. The algorithm should overcome the problem of scalability.
8. The generated clusters should facilitate hierarchical routing.

10. PROPOSED ALGORITHM – MODIFIED DSECA (M_DSECA)

The proposed algorithm is a modified version of DSECA given in [13]. The algorithm is an extension of the 3hBAC clustering algorithm [14] and the weight-based clustering algorithms. As in LCC, 3hBAC and other clustering algorithms, the proposed clustering procedure also involves two phases, namely, cluster formation phase and cluster maintenance phase.

10.1. Notations used in M_DSECA

* The tuple **(m, p)** denotes *a master and its corresponding proxy pair.*

* **(M, P)** denotes *the set of all (m, p) such that m is a master and p is its corresponding proxy pair.*

* **hm-I** denotes *a node which is a hidden master of type I.*

* **HM-I** denotes *the set of those nodes which are hidden masters of type I.*

* **hm-II** denotes *a node which is a hidden master of type II.*

* **HM-II** denotes *the set of those nodes which are hidden masters of type II.*

* **N(u)** denotes *the set of all 1-hop neighbours of u.*

* **N'(u)** denotes *the set of those neighbours of u having greater weight than u.*

* **N''(u)** denotes *the set of those neighbours of u which are not Master/Proxy nodes and also having lesser weight than that of u.*

* $N_m(u)$ denotes *the set of those neighbours of u which are adjacent to some Master node.*

10.2. Cluster set up phase

M_DSEC(*G*).

In the cluster set up phase, initially, all the nodes are grouped into some clusters.

Initial (Master, Proxy) election. Among all the nodes in the network, choose a node having maximum weight. It is designated as a Master. Next, among all its neighbors, the one with greater weight is chosen and it is designated as a Proxy. Then the initial cluster is formed with the chosen Master, Proxy and their neighbors. Since this structure will embed in itself a double star, the algorithm is referred to as a double star embedded clustering algorithm.

Second and subsequent (Master, Proxy) election. For the subsequent (Master, Proxy) elections, we impose an additional condition on the hop distance between different (m, p) pairs to generate non-overlapping clusters. Here, as in 3hBAC, we impose the condition that all the (m, p) pairs should be atleast 3 hops away from each other. Nodes which are already grouped into some clusters are excluded in the future cluster formation processes. Among the remaining pool of nodes, choose the one with higher weight. Next,

(i)   Check whether the newly chosen node is exactly 3-hop away from atleast one of the previously elected Masters (or Proxies) and atleast 3-hop from the corresponding Proxy (or Master)

(ii) The newly elected node should be at distance atleast 3-hop from rest of the (m, p) pairs.

If the above chosen higher weight node satisfies these two conditions, it can be designated as a Master.

To choose the corresponding Proxy, among the neighbours of above chosen Master, find the one with higher weight and at distance atleast 3-hop from each of the previously elected (m, p) pairs. Then, obtain a new cluster with this chosen (m, p) pair and their neighbours. Repeat this procedure until all the nodes are exhausted. The nodes which are not grouped into any cluster and the set HM-I are collected separately and termed as "Critical nodes". The set of all nodes grouped into some clusters is denoted by S and the set of critical nodes is denoted by C. Hence, after the cluster set up phase, the sets S and C are obtained as output. The pseudo code for the above process is given below.

10.2.1. Main Procedure

M_DSEC(G)

1. Randomly generate the required node positions of all the nodes in the network.
2. **for** each node $u \in N$, compute N(u)
3. Compute the Euclidean distance matrix and Hop distance matrix.
4. S=3-hop-M_DSEC(N) /*Calling procedure to form a set with maximum possible 3-hop perfect double star embedded clusters*/
5. C = N/S ∪ CR   // C forms the Critical node set
6. **If**(C == ϕ)
7.   **then**
            Print "Perfect clustering…" & goto step 15
8.   **else** {
9. S_A = adjusted_M_DSEC(S, C)
10. C_A = N\S$_A$
11.   **Print** "Refined Clustering…"
12.   **Return** S_A, C_A
13.   **Exit**
14. }**Endif**
15. **Return S, C**
16. **Exit**

10.2.2. Formation of perfect 3-hop modified Double star embedded clusters

3-hop-M_DSEC(N)

1. **for** each vertex $u \in N$, compute W(u)
2. S ← ϕ //Union of all double star embedded clusters

3.   CR $\leftarrow \phi$ //Union of hidden masters of type I

4.   j $\leftarrow$ 1

5.   Extract a node, say x, from N such that W(x) is maximum. (In case of a tie, choose the one with higher NS value)

6.   Find N(x)//Nodes within one hop from x

7.   From N(x), extract a node with maximum weight. Label it as y.

8.   Find N(y) and N'(y) = set of those neighbors of y having greater weight than y

9. $C_j \leftarrow$ {x, y}$\cup$N(x)$\cup$N(y)   /* Initial cluster formation, x acts as master, y acts as proxy and its neighbors are slaves */

10.   Master[j]$\leftarrow$x, Proxy[j]$\leftarrow$y, HM[j]$\leftarrow$ N'(y)

11.   S$\leftarrow$S$\cup$$C_j$ //Updation of double star embedded clusters

12.   CR = CR$\cup$HM[j]

13.   P=ϕ /*Set of those nodes with higher weight but not eligible for Master because of not satisfying distance property */

14.   **do{**

15.   Extract a node, say x, from N\[S$\cup$P] such that W(x) is maximum. (In case of a tie, choose the one with higher NS value). Label the newly chosen node as z.

16.   **If**((d(z, Master[i])==3) && d(z, Proxy[i]$\geq$3)) ||

      (d(z, Proxy[i]==3) && d(z, Master[i]$\geq$3)), for some 1$\leq$i$\leq$j) {

17.   **If**((d(z, Master[k])$\geq$3&&d(z, Proxy[k])$\geq$3),

       for all k$\neq$i and 1$\leq$ i, k$\leq$j) {

18.   j$\leftarrow$j+1

19.   x $\leftarrow$ z

20.   Goto step 6

21.   }

22.   **else {**

23.   P = P$\cup$z

24.   Goto step 14

25.   }**Endif**

26. **else {**

27.   P = P$\cup$z

28.   Goto step 14

29. }**Endif**

30. **}while**(N\[S$\cup$P]$\neq$ ϕ)

31. **Return** S, CR

32. **Exit**

10.3. Cluster Maintenance phase – Treatment of critical nodes generated by M_DSECA

10.3.1. Nature of Critical nodes

A node in the critical node set C generated after implementing M_DSECA may be of any one of the following categories:
(i) A Hidden Master of type I.
(ii) A Hidden Master of type II.
(iii) A Node neglected in cluster formation because of lesser weight.

10.3.2. Neighbors of critical nodes

Let u be a critical node and v be a neighbor of u. Then the following cases may arise:

Case i: v is another critical node
Subcase i: v is a hm-I.
Then v will have an adjacent Proxy node such that w(v) is greater than that of the proxy node. In this case, the set N''(v) will be used to form adjusted clusters.
Subcase i: v is not a hm-I.
In this case, the set $N''(v)\backslash N_m(v)$ will be used for adjusted cluster formation.

Case ii: v is a slave node (an existing cluster member)
In this case, v will be used for adjusted cluster formation provided v is not adjacent to any existing Master. In such a case, we consider all the neighbors of v having lesser weight than v except the neighbors which are Proxy nodes.
If v is adjacent to some master, then it will not be used for adjusted cluster formation.

10.3.3. Formation of adjusted Double star embedded clusters

In the formation of adjusted clusters, we try to form clusters of these critical nodes either among themselves or by extracting nodes from existing clusters and regroup them with critical nodes to form better clusters. The below adjusted clustering procedure is invoked to minimize the number of critical nodes.

adjusted_M_DSEC(S, C)

From C, extract a node with maximum weight. Let it be c. Then any one of the following cases arises. Here, we form the adjusted clusters depending on the nature of the critical nodes by considering only restricted neighbours as explained in section 10.3.2.

Case i: c is a hm-I. Then, c will have an adjacent Proxy, say p. From $N(c)\backslash\{p\}$, choose a node, say c', having greater weight.

Subcase i: c' is another critical node. If c' is a hm-I, then find N''(c') and form the new adjusted double star embedded cluster with $\{c, c'\} \cup (N(c)\backslash\{p\}) \cup N''(c')$. The node c acts as the Master and c' as the Proxy of the new adjusted cluster. Otherwise, the set {c, c'} $\cup (N(c)\backslash\{p\}) \cup N(c')$ will form the new adjusted double star embedded cluster with c as Master and c' as Proxy.

Subcase ii: c' is a slave node. In this case, c' may be adjacent to some Master/Proxy of existing clusters. As explained in 10.3.2., if c' is adjacent to any master, then it will not be used for adjusted cluster formation. If not, then we obtain a new adjusted cluster with $\{c, c'\} \cup (N(c)\backslash\{p\}) \cup N''(c')$.

Case ii: c is not a hm-I. In this case, from $N''(c)\backslash N_m(c)$, choose a node, say c' having greater weight, then form a new adjusted cluster, with c, c', $N''(c)\backslash N_m(c)$ and $N''(c')\backslash N_m(c')$.

Repeat this procedure until either all the nodes are exhausted or no such selection can be performed further. If there is any node still left uncovered after completing this procedure, it will become a Master on its own.

Further, as the position of nodes may change frequently due to mobility, each (m, p) pair should periodically update its neighbor list so that if any slave node moves outside its cluster boundary, it can attach itself to its neighboring cluster by passing find_CH messages to all (m, p) pairs. If it receives an acknowledgment from some Master/Proxy, it will join that cluster. In case getting an acknowledgement from two or more nodes, the slave chooses the one with higher weight.

11. AN ILLUSTRATION

The above given procedure is explained with the following network graph. Consider the network graph shown in Figure 4., consisting of 23 nodes. The (x, y) positions of the nodes in the network are randomly generated and the graph is plotted with those positions. The weight of each node is computed using formula 1. Appendix gives a detailed description of the computation of weights of the nodes in the below network graph. The values in the parentheses denote the node id and weight values of the respective nodes. Here, the value of NS(u) is computed by arbitrarily considering some of the neighbours as strong, some as weak and some as medium neighbours and taking the threshold value K=100.

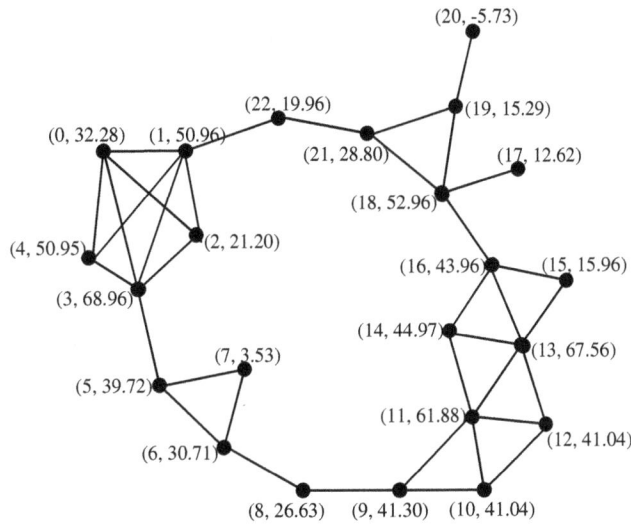

Figure 4. An Example Network Graph (G)

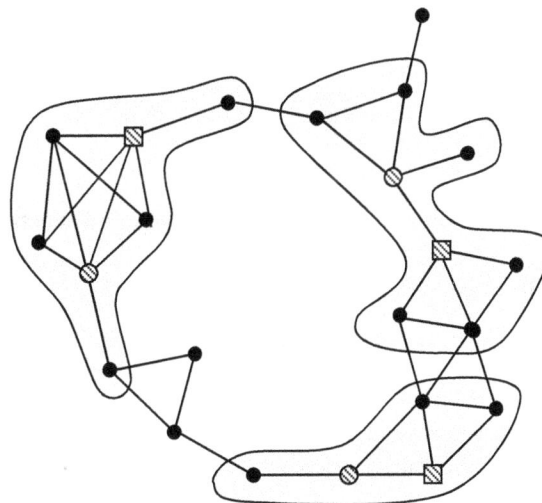

Figure 5. Clusters generated after executing **M_DSEC(G)**

(Here, the striped circles denote Masters, Striped squares denote Proxies and Shaded circles denote slaves)

Clusters generated after executing M_DSEC(G):

$C_1 = \{(\mathbf{3, 1}), 0, 2, 4, 5, 22\}$
$C_2 = \{(\mathbf{18, 16}), 13, 14, 15, 17, 19, 21\}$
$C_3 = \{(\mathbf{9, 10}), 8, 11, 12\}$

After executing **M_DSEC(G)**, the adjustment procedure **adjusted_M_DSEC(S, C)** explained in section 10.3.3. is executed with $S = C_1 \cup C_2 \cup C_3$ and HM-I = {11, 13, 14}, HM-II = {11, 13}, C = HM-I $\cup$ Set of nodes left unclustered = {11, 13, 14, 6, 7, 20}. Among the nodes in C, the node 13 possesses highest weight. Hence, it becomes an eligible master for adjusted cluster formation. Now, node 13 is a hm-I. Therefore, it has an adjacent proxy, i.e., node 16. Hence, by looking into N(13)\{16}, we choose a node with higher weight, which node 11. Thus, by using case (i) of section 10.3.3., we get a new adjusted cluster $C_1' = \{(\mathbf{13, 11}), 12, 14, 15\}$. At the same time cluster C_2 gets changed as $C_2 = \{(\mathbf{18, 16}), 17, 19, 21\}$. Then continuing with the remaining set of critical nodes, i.e., {6, 7, 20}, we get another new adjusted cluster $C_2' = \{6, 7\}$. The node 20 is still left uncovered. So, it is declared as a master on its own. Thus, the adjusted clusters obtained finally will be as shown in Figure 6. It can be seen from Figure 5 and Figure 6 that after implementing the adjustment procedure, not only the critical nodes are grouped into some clusters, but also the already generated clusters get adjusted automatically so that the load is well balanced. Hence, the algorithm generates optimum load balancing clusters.

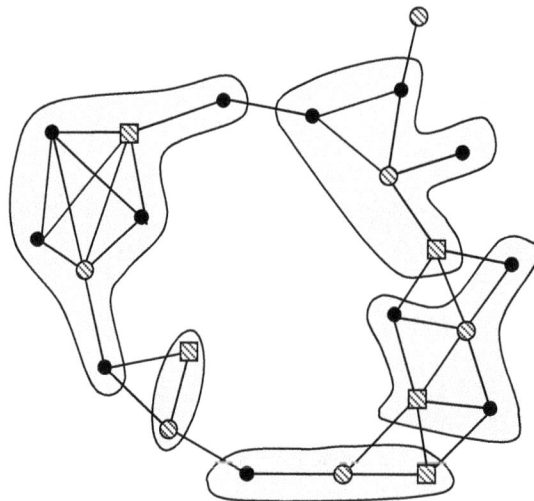

Figure 6. Adjusted Clusters formed after executing **adjusted_M_DSEC**

12. CATEGORIZATION OF M_DSE CLUSTERING

Any clustering which yields no critical nodes after initial cluster formation is said to be a *perfect clustering.* The one which yields some critical nodes but the number can be reduced to zero after the execution of adjustment procedure given in 9.2.3. is said to be a *fairly-perfect clustering* scheme and the one in which the number of critical nodes cannot be reduced to zero even after implementing the adjustment procedure is said to be an *imperfect clustering.*

13. PROPERTIES OF THE CLUSTER STRUCTURES

In general, to meet the requirements of the ad hoc networks, a clustering algorithm is required to partition the nodes of the network so that three ad hoc clustering properties are satisfied. (1) Dominance Property, (2) Independence Property, (3) Guaranteed good service by the leader nodes.[3, 21]. It can be seen that the proposed algorithm also satisfies the above properties. i.e.,

1. Every ordinary node (a node which is neither a master nor a proxy) affiliates with a leader node (Master/Proxy) (dominance property).

2. As per the proposed algorithm, the Master nodes are always maintained to have higher weight than the rest of the nodes in that cluster.

3. Every slave node is at most d hops away from its (Master, Proxy)-pairs, where $d = 2$.

4. No two Master nodes are adjacent (guarantees well scattered clusters).

The following are some of the other graph theoretic/structural properties observed in the cluster structures obtained using the proposed algorithm.

Property 1. Each cluster is of diameter atmost 3.

Property 2. Each double star embedded subgraph has a dominating edge

Property 3. After finishing the execution of both M_DSEC and the adjustment procedure, each vertex lies in exactly one cluster, as each slave node is affiliated with exactly one (Master, Proxy)-pair, whereas each critical node is declared itself as a leader node.

Property 4. If the resultant clustering is a *perfect clustering,* then the set of all (Master, Proxy)-pairs will form an *efficient edge dominating set* and the total number of clusters obtained in such a case will be equal to the domination number of the line graph of the underlying network graph.

14. CONCLUSION AND FUTURE WORK

The proposed algorithm yields a cluster structure, where the clusters are managed by Master nodes. In case of any failure of Master nodes, the cluster is not disturbed and the functions are handed over to an alternative which behave in a similar way to Master nodes (Perhaps with less efficiency than Masters but better than ordinary nodes). In order to better suite practical constraints, we have included the Euclidean-closeness measures in addition with the hop-closeness measure used in [13]. This enables us to increase the life time of the network. Further, since the clusters are managed by the Masters as well as by the Proxy nodes at times of necessity, the load is well balanced. The event of re-clustering can be avoided as long as possible. The algorithm is being implemented in NS2 and the expected better performance of the algorithm will be guaranteed on comparison of this with the other existing algorithms.

REFERENCES

[1] C.E. Perkins, "Ad hoc Networking", Addison Wesley, Pearson Education, Inc. And Dorling Kindersley Publishing, Inc., India © 2001.

[2] L. Ramachandran, M. Kapoor, A. Sarkar and A. Aggarwal, "Clustering Algorithms for wireless ad hoc networks", Proc. of 4th International Workshop on Discrete algorithms and methods for mobile computing and communications, Boston, MA, August 2000, pp. 54-63.

[3] S. Basagni, "Distributed Clustering for Ad hoc networks", Proc. of ISPAM'99 International Symposium on parallel architectures, algorithms and networks, pp. 310-315, 1999.

[4]   M. Chatterjee, S.K. Das and D. Turgut, "A Weight based distributed clustering algorithm for MANET", V.K. Prasanna, et. al (eds.) HiPC 2000, LNCS, vol. 1970, pp. 511-521, Springer, Heidelberg (2000).

[5]   M. Chatterjee, S.K. Das and D. Turgut, "WCA: A Weighted Clustering Algorithm for Mobile Ad Hoc Networks", Cluster Computing, vol. 5, pp. 193-204, Kluwer Academic Publishers, The Netherlands, 2002.

[6]   W. Choi and M. Woo, "A Distributed Weighted Clustering algorithm for mobile ad hoc networks", Proc. of AICT/ICIW 2006, IEEE, Los Alamitos, 2006.

[7]   W. Yang and G. Zhang, "A Weight-based clustering algorithm for mobile ad hoc networks", Proc. of 3rd International Conference on Wireless and mobile communications, 2007.

[8]   I.Y. Kim, Y.S. Kim and K.C. Kim, "Zone-based clustering for intrusion detection architecture in ad hoc networks", Management of Convergence networks and services, APNOMS 2006 Proceedings, LNCS, vol. 4238, pp. 253-262, Springer, Heidelberg, 2006.

[9]   Y.P. Chen, A.L. Liestman and J. Liu, "Clustering Algorithms for ad hoc wireless networks", in Ad hoc and Sensor Networks, Y. Pan and Y. Xiao (eds.), Nova Science Publishers, pp. 1-16, 2004.

[10]   K. Erciyes, O. Dagdeviren, D. Cokuslu and D. Ozsoyeller, "Graph Theoretic clustering algorithms in mobile ad hoc networks and wireless sensor networks Survey", Appl. Comput. Math. Vol. 6, No. 2, pp. 162-180, 2007.

[11]   J.Y. Yu and P.H.J. Chong, "A survey of clustering schemes for mobile ad hoc networks", First Quarter, Vol. 7, No. 1, pp. 32-47, 2005.

[12]   S.J. Francis, E.B. Rajsingh, "Performance analysis of clustering protocols in mobile ad hoc networks", J. Computer Science, Vol. 4, No. 3, pp. 192-204, 2008.

[13]   T.N. Janakiraman and A.S. Thilak, "A Weight based double star embedded clustering of homogeneous mobile ad hoc networks using graph theory", Advances in Networks and Communications, N. Meghanathan et al. (eds.), CCIS, Vol. 132, Part – II, pp. 329-339, Proc. of CCSIT 2011, Springer, Heidelberg, 2011.

[14]   J.Y. Yu and P.H.J. Chong, "3hBAC (3-hop Between Adjacent cluster heads: a novel non-overlapping clustering algorithm for mobile ad hoc networks", Proc. of IEEE Pacrim 2003, Vol. 1, pp. 318-321, 2003.

[15]   C. Chiang, "Routing in clustered multihop, mobile wireless networks with fading channel", Proc. of IEEE SICON '97, 1997.

[16]   A.A. Abbasi, M.I. Buhari and M.A. Badhusha, "Clustering Heuristics in wireless networks: A survey", Proc. 20th European Conference on Modelling and Simulation, 2006.

[17]   D.J. Baker and A. Epremides, "A distributed algorithm for organizing mobile radio telecommunication networks", Proc. 2nd International conference on distributed computer systems, pp. 476-483, IEEE Press, France, 1981.

[18]   M. Gerla and J.T.C. Tsai, "Multi-cluster, mobile, multimedia radio network", Wireless networks, vol. 1, No. 3, pp. 255-265, 1995.

[19]   A.K. Parekh, "Selecting routers in ad hoc wireless networks", Proc. SB/IEEE International Telecommunications Symposium, IEEE, Los Alamitos, 1994.

[20]   P. Basu, N. Khan, T.D.C. Little. "A mobility based metric for clustering in mobile ad hoc networks", Proc. IEEE ICDCS, Phoenix, Arizona, USA, pp. 413-418, 2001.

[21]   E.R. Inn and W.K.G. Seah, "Performance analysis of mobility-based d-hop (MobDHop) clustering algorithm for mobile ad hoc networks", Computer Networks, vol. 50, 3339-3375, 2006.

[22]   T.W. Haynes, S.T. Hedetniemi and P.J. Slater, "Fundamentals of Domination in Graphs", Marcel Dekker, Inc., New York.

12

OPPORTUNISTIC AND PLAYBACK-SENSITIVE SCHEDULING FOR VIDEO STREAMING

Huda Adibah Mohd Ramli[1] and Kumbesan Sandrasegaran[2]

[1]Department of Electrical and Computer Engineering, International Islamic University Malaysia (IIUM), Kuala Lumpur, Malaysia

[2]Faculty of Engineering and Information Technology, University of Technology, Sydney, Australia

ABSTRACT

Given the strict Quality of Service (QoS) requirements of video streaming, this paper proposes a novel solution for simultaneous streaming of multiple video sessions over a mobile cellular system. The proposed solution combines a buffer management strategy with a packet scheduling algorithm. The buffer management strategy selectively discards packets of a user from base station buffer whereas the packet scheduling algorithm schedules packets of a user according to its instantaneous channel quality, average throughput and playback buffer information. Simulation results demonstrate that the proposed solution is effective in providing a continuous video playback with good perceptual quality for more users. If at least a good perceptual quality is to be satisfied for all users (QoS constraint of video streaming), then the proposed solution improves the system capacity by 40% over a conventional packet scheduling algorithm.

KEYWORDS

Packet scheduling, video streaming, Long-Term Evolution, Quality of Service, Orthogonal Frequency Division Multiple Access.

1. INTRODUCTION

Long Term Evolution (LTE), which is now referred to as 3.9G, is the latest commercially available Third Generation Partnership Project (3GPP) standard. The LTE is envisaged to provide a better quality of multimedia communications by providing higher data rates (50 Mbps in uplink and 100 Mbps in downlink), reduced latency and increased capacity and coverage. The LTE is a multi-carrier mobile cellular system. It uses Orthogonal Frequency Division Multiple Access (OFDMA) for downlink transmission. The bandwidth in the downlink LTE is divided into multiple equally spaced and mutually orthogonal sub-carriers. The minimum downlink LTE transmission unit that can be allocated to a user is referred to as a Resource Block (RB). An RB is made up of 12 sub-carriers of 1 ms duration [1].

Recent trends have shown an increase in popularity of video streaming application among mobile cellular users [2]. For transport over mobile cellular channels, a video stream is encoded (compressed) into frames of different properties (namely I, P and B frames) in order to reduce the bandwidth requirements [3]. There are multiple Group of Pictures (GoPs) within an encoded video stream. A GoP starts with an I frame and all frames prior to the subsequent I frame [4]. Each frame within a GoP has a different priority and is highly co-related. The I frame has the highest priority followed by the P and B frames. It should be noted that decoding of a P frame within a GoP is dependent upon an I frame within the same GoP and decoding of a B frame within a GoP is dependent upon an I and P frames within the same GoP. The loss of a higher priority frame within a GoP results in the loss of other dependent frames within the same GoP [5]. Even when the video stream is compressed, it still requires large bandwidth. Simultaneous transmissions of this bandwidth-hungry video streaming application is challenging as it may lead to a mobile cellular congestion if the expensive radio resources are not properly scheduled. Another challenge is that streaming of multiple video sessions simultaneously requires a continuous video playback at the highest perceptual quality at each user (Quality of Service, QoS, constraint of video streaming) [6].

Numerous studies have been discussed in the literature so as to address the stated challenges. For example, the authors in [7] developed a buffer management strategy that selectively discard packets of video users at the base station according to their priority and deadline. Note that packet is a segment of a video frame. This strategy ensures that the limited radio resources are efficiently used for transmission of packets that can be used for decoding and video playback and hence improving the perceptual quality experience at the users. Frequent interruptions during video playback may cause major annoyance to the video users. As such, a buffer management strategy that attempts to improve video playback continuity by minimizing the number of interruptions during video playback was developed in [8].

Besides the buffer management strategy, packet scheduling is another area of research interest when dealing with simultaneous transmission of multiple video users. Packet scheduling is responsible to efficiently select a user's packets for (re)transmission at a given time using an available radio resource so as to provide a satisfactory QoS, guarantee fairness and optimize system performance. Proportional Fair (PF) [9] is one of the well-known packet scheduling algorithms in the legacy single-carrier mobile cellular systems. This algorithm takes channel quality and average throughput of each user into consideration when making scheduling decision. In addition to that, a packet scheduling algorithm that delays transmission of the least important packets and allocates more radio resources for transmission to more important packets was developed in [10]. The algorithm was developed in order to avoid mobile cellular congestion and ensure a continuous video playback.

Other studies in the literature combined a buffer management strategy with packet scheduling algorithm. It should be noted that the following discussions refers the combination of a buffer management strategy with packet scheduling algorithm as a packet scheduling solution. To ensure timely arrival of video packets at a user, the authors in [11] developed a packet scheduling algorithm that schedules packets of a user on the basis of its channel quality and frame delay. This packet scheduling algorithm is then combined with a buffer management strategy that discards packet and other dependent packets if they are likely to arrive at the user end after the playback deadline. The playback deadline is the time when a frame is needed for video playback and each frame is attributed with this deadline. A packet scheduling solution that aimed to improve the perceptual quality among all video users was developed in [12]. The developed packet scheduling algorithm schedules packets of a user according to its perceptual quality, channel quality and decoding deadline. Decoding deadline is the time when a frame is needed for decoding at a user end. The buffer management strategy developed in this study discards a packet and other dependent packets at base station if the time when the packet is needed for decoding at a user end has passed. Similarly, the authors in [13] developed a solution that attempts to improve perceptual quality across all video users. The packet scheduling algorithm developed in the study determines priority of a user's packets according to its channel quality, average throughput, priority of a frame, playback buffer information (the playback buffer is located at Application Layer at a user end) and whether any packets have been transmitted to the user or not. If a packet is discarded at a user for playback deadline violation, the developed buffer management strategy discards all packets that are dependent to the discarded packet at the base station.

The aforementioned studies mostly focus in satisfying either the perceptual quality or video playback continuity, but not both. A continuous video playback does not mean that an excellent perceptual quality is experienced at the user (i.e. the video playback may be continuous but with a poor perceptual quality as a number of frames in a video stream is lost). Similarly, a user may experience a good perceptual quality but has to tolerate with a large number of playback interruptions (i.e. a video playback is interrupted if a frame is not available when it is needed for playback or due to playback buffer underflow). To address this situation, this paper proposes a novel packet scheduling solution known as Opportunistic and Playback-Sensitive Scheduling (OPSS) for usage in the downlink LTE. OPSS is a combination of a buffer management strategy and a packet scheduling algorithm. It aims to optimize the system capacity (number of users) without compromising the QoS constraint of video streaming (i.e. a continuous video playback at the highest perceptual quality for each user).

The remaining sections of this paper are organized as follows: Section 2 gives an overview of the video streaming followed by a detailed description of the proposed packet scheduling solution in Section 3. Section 4 contains environment of the simulation while Section 5 evaluates performance of the OPSS against a well-known packet scheduling algorithm. Finally, Section 6 concludes the paper.

2. VIDEO STREAMING OVERVIEW

As previously discussed in Section 1, simultaneous streaming of multiple video sessions requires a continuous video playback at the highest perceptual quality at each user. Generally, the perceptual quality is measured on the basis of Peak-Signal-to-Noise-Ratio (PSNR) [14] or Mean Opinion Score (MOS) [15] metrics. Note that the PSNR of a user can be mapped to a MOS value to give a qualitative representation of the perceptual quality. The MOS contains a scaled value from 1 to 5 that implies a bad to an excellent perceptual quality (as illustrated in Table 1).

Table 1. Average PSNR to MOS mapping [16]

Lowest Average PSNR (dB)	Minimum MOS	Perceptual Quality
< 20	1	Bad
20 – 25	2	Poor
25 – 31	3	Fair
31 – 37	4	Good
> 37	5	Excellent

Video streaming allows a user to start a video playback without receiving an entire video stream. A typical approach for ensuring a continuous playback is to delay the start of a video playback until the total number of frames in the playback buffer exceeds a buffering threshold [17]. The buffering threshold is defined as the maximum number of frames required to be filled in the playback buffer at a user end before the user can start or resume a video playback.

The video streaming properties and the mobile cellular channels characteristics may lead to interruptions during the video playback [18, 19]. Interruption occurs when the total number of frames in the playback buffer is less than a playback buffer underflow threshold. The playback buffer underflow threshold is defined as the maximum number of frames required to be available in the playback buffer for a continuous playback. The video playback resumes at the same position where the interruption occurs after the total number of frames in the playback buffer exceeds the specified buffering threshold.

A user will not be satisfied if it has to wait longer to start or resume its video playback. Freezing Delay Ratio (FDR) metric [6] which gives a total delay experienced by each user throughout its video session is used to measure the video playback continuity. Note that a low FDR implies that all users are likely to experience continuous video playback throughout their sessions. A high FDR indicates that a number of users are likely to have waited for an amount of time before they can start their video playback or there a number of interruptions occurred throughout these users' video sessions.

An example of the freezing delay where it takes ε ms for the first bit of the video stream to arrive and being stored in the playback buffer after a user requests for a video session is illustrated in Figure 1. The video playback starts after it has been delayed for α ms. This delay is used to fill in the playback buffer such that the total number of frames in the playback buffer exceeds the buffering threshold. The video playback is interrupted if the total number of frames is less than the playback buffer underflow threshold. The video playback resumes after β ms. This time duration is used to re-fill in the playback buffer such that the total number of frames in the playback buffer exceeds the buffering threshold.

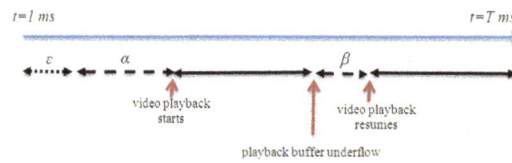

Figure 1. Example of freezing delay during a video session

A frame needs to be decoded first before it can be played back. Some frames cannot be decoded if a higher priority frame is lost. These frames are discarded at the user. A number of studies (i.e. [7, 11, 12]) considered a scenario that discards a frame and other dependent frames if the frame arrives at the playback buffer after its decoding or playback deadline. These studies generally aimed to minimize frame loss ratio as the loss of frames due to frame discards degrades the perceptual quality experienced at the users. However, this paper considers a scenario that does not discard packets/frames for decoding/playback deadline violation.

3. OPPORTUNISTIC AND PLAYBACK SENSITIVE SCHEDULING

This section presents a novel packet scheduling solution called OPSS so as to ensure a continuous video playback at the highest perceptual quality for more users. This solution is proposed for usage in the multi-carrier downlink LTE. Detailed descriptions of the buffer management strategy and packet scheduling algorithm of the OPSS are discussed in the following sections.

3.1 Buffer Management Strategy

All packets arriving at the user end are delivered in sequence towards the playback buffer through the use of re-sequencing buffer. Whenever a TB (with a known Transmission Sequence Number - TSN) is erroneously received at the user, all subsequent TBs with higher TSNs that are correctly received by the user are stored in the re-sequencing buffer and a re-sequencing timer is started. The packets of out-of-sequence TBs are delivered towards the playback buffer upon a correct reception of the erroneous TB or upon expiry of the re-sequencing timer [20]. In this paper, packets of a TB are discarded and considered as lost packets only if the TB has exceeded maximum number of retransmissions or the re-sequencing timer associated with the TB has expired.

The loss of video packets will degrade video quality. Degradation will be significant if higher priority packets that are to be used for decoding other dependent packets are lost. Whenever a packet is lost, there may be one or more packets belonging to the same frame to the lost packet or are dependent to the lost packet are still residing in the base station buffer. These packets should not be transmitted to the user as the packets cannot be used for decoding or video playback (due to their dependency to the lost packet). Therefore, the buffer management strategy of the OPSS discards these packets from the base station buffer such that the available radio resources are efficiently used for (re)transmission of packets that can be used for decoding and video playback. All packets are stored in a transmission buffer at the base station upon transmission. The time duration that each packet has been residing in the transmission buffer is known to the base station. Moreover, the maximum duration that a packet can reside in a re-sequencing buffer at the user is also known at the base station. One or more packets may have resided long enough within the transmission buffer such that, if they are retransmitted, they are likely to arrive at the user end after the re-sequencing timer expires. To avoid this situation, the buffer management strategy discards these packets from the transmission buffer.

The buffer management strategy proposed in this paper is almost similar to the strategies discussed in [7, 11-13] that discard packets at the base station (i) due to packets dependency or (ii) if the packets are likely to arrive at the user after the deadline. Additionally, similar to [13], the playback buffer information is considered in the proposed packet scheduling algorithm as it plays a role in ensuring a continuous video playback [19]. Note that the playback buffer is only updated when all packets belonging to a frame have correctly arrived at the playback buffer. On the contrary, the OPSS differs from other solutions as the proposed packet scheduling algorithm uses different formulation when determining priority of each user for packets transmission. Detailed description of the proposed packet scheduling algorithm is described next.

3.2 Packet Scheduling Algorithm

In each scheduling interval and on each RB, the packet scheduling algorithm of the OPSS schedules a user that maximizes $\mu_{i,j}(t)$ in the following equation:

$$\mu_{i,j}(t) = \left(\frac{r_{i,j}(t)}{\left(R_i(t)\right)^2}\right) * \left(1 - \frac{PB_i(t)}{\alpha_{pb}}\right) \tag{1}$$

$$R_i(t+1) = \left(1 - \frac{1}{t_c}\right)R_i(t) + \frac{1}{t_c} * rtot_i(t+1) \tag{2}$$

$$rtot_i(t+1) = \sum_{j=1}^{RB_{max}} r_{i,j}(t+1) \tag{3}$$

where $\mu_{i,j}(t)$ is the priority of user i on RB j at scheduling interval t, $r_{i,j}(t)$ is the instantaneous data rate of user i on RB j at scheduling interval t, $R_i(t)$ is the average throughput of user i at scheduling interval t, $PB_i(t)$ is the total number of frames in playback buffer of user i at scheduling interval t, α_{pb} is the playback buffer weighting factor (it scales the second multiplicand in Equation (1) to be between 0 and 1), t_c is a time constant, $rtot_i(t+1)$ is the total data rate being used to transmit packets to user i at scheduling interval $t+1$, and RB_{max} is the maximum available number of RBs. Note that the playback buffer weighting factor is used to avoid the scheduling decision to be dependent upon playback buffer information only as the value for the first multiplicand in Equation (1) is between 0 and 1.

The proposed algorithm is more likely to give scheduling opportunity to a user with the least number of frames in its playback buffer if the channel quality and the average throughput of each user are similar. If one or more TBs are erroneously received at the user, it will take a longer time to update the playback buffer as all subsequent TBs with higher TSNs than the erroneous TBs that may have been correctly arrived at the user are stored in the re-sequencing buffer. The packets of out-of-sequence TBs are only delivered to the playback buffer upon correct reception of the erroneous TBs or upon expiry of the re-sequencing timer. Scheduling opportunity is highly likely to be given to this user in subsequent scheduling intervals if its playback buffer is not updated. The average throughput is used in the proposed algorithm to compensate for the effect of the delay in playback buffer update. In this case, the user priority decreases in subsequent scheduling intervals as more packets are transmitted to the user. After a number of scheduling intervals, even if the user has the least number of frames in its playback buffer, scheduling opportunity is highly likely to be given to other users due to the transmission history of this user.

It can be observed in Equation (1) that the formulation of the proposed packet scheduling algorithm is dependent upon three variables namely channel quality, average throughput and playback buffer information. The relevance of each variable is justified next. A number of studies (i.e. [21-24]) have shown that scheduling opportunity that considers channel quality can significantly improve the system performance. Therefore, this variable is considered in the proposed algorithm. However, even though the packets of each selected user can be transmitted with a better modulation and coding scheme, the packet scheduling algorithm that is dependent upon channel quality alone is inefficient in fairness as it deprives users located at the cell edge from receiving their packets.

To address the fairness problem, the average throughput variable is taken into consideration in the proposed algorithm. The update of the average throughput increases the priority of users whose packets are not scheduled in the previous scheduling intervals. Therefore, the packets of these users are likely to be scheduled in the sub-sequent scheduling intervals. Additionally, this variable is used to compensate for the effect of delay in playback buffer update (as discussed previously in this section). Finally, the playback buffer information is considered in order to ensure that all users are given an equal opportunity for video playback. This allows the OPSS to minimize the delay before the users can start or resume their video playback. Moreover, this variable attempts to maintain a continuous video playback by giving a higher priority for packets transmission to the users with least number of frames in their playback buffer (if the channel quality and the average throughput of each user are similar) so as to avoid the video playback from being interrupted due to playback buffer underflow.

4. SIMULATION ENVIRONMENT

The performance of the OPSS was evaluated within a single hexagonal cell scenario of 5 MHz bandwidth with 25 RBs and 2 GHz carrier frequency. The base station has a fixed location at the centre cell and it was assumed that equal transmit power (43.01 dBm total base station transmit power) is used on each RB. Each user moves at a constant speed of 3 km/h. These users are uniformly located within the cell. The Cost-231 HATA model for an urban environment [25], a Gaussian log-normal distribution with 0 mean and 8 dB standard deviation [26] and a frequency-flat Rayleigh fading model [27] are used to model the radio propagation channel. The probability that the channel quality information report is in error was fixed at 1% and this report is only available for use by the base station after a 4 ms delay [28]. It was assumed that besides the CQI and Hybrid Automatic Repeat Request (HARQ) information, the feedback from a user also contains playback buffer information.

Three video streams which were downloaded from a publicly available video traces [4] were used. The video streams were encoded with a frame rate of 30 fps (frames per second). One GoP contains 16 frames with IBBBPBBBPBBBPBBB sequence. Each frame has a number, type (I, P or B frame), decoding/playback deadline, size and PSNR value of the luminance component. A user randomly requests one video stream throughout its session and the request arrives at the base station at the beginning of the simulation. It was assumed in this performance study that: (i) a video playback is interrupted if a frame is not available when it is needed for playback (i.e. the frame is not discarded if it arrives at the playback buffer after the playback deadline [18, 19]), (ii) the playback buffer capacity of each user is infinite and (iii) the playback buffer underflow threshold is fixed at 5 frames.

Minimum MOS and FDR metrics are used for performance evaluation. In this paper, the minimum MOS evaluates the perceptual quality where a user with the lowest average PSNR among other video users is considered. In this case, a low minimum MOS (i.e. minimum MOS=1, see Table 1) indicates that at least one user is experiencing a bad perceptual quality whereas a high minimum MOS (i.e. minimum MOS=4) indicates that all users are experiencing at least a good perceptual quality. The FDR is defined in this paper as the average of the ratio of total freezing delay of each user to the total simulation time. Equation (4) gives the mathematical expression of the FDR [29]:

$$FDR = \frac{1}{N} \sum_{i=1}^{i=N} \frac{Df_i}{T} \qquad (4)$$

where Df_i is the total freezing delay of user i, T is the total simulation time and N is the total number of users.

The results obtained via computer simulation of the OPSS solution were evaluated and compared with the well-known PF algorithm. The PF was chosen as it is one of the packet scheduling algorithms that provides considerably good performance in the legacy single-carrier mobile cellular systems. Moreover, due to its efficiency, the authors in [30-33] all extended the PF algorithm into multi-carrier mobile cellular systems. Packet scheduling in the single-carrier mobile cellular systems allocates all of the available radio resources to a single user in each scheduling interval. As such, the PF equations that support packet scheduling in the downlink LTE system as described [33] is used in this performance evaluation.

5. PERFORMANCE RESULTS AND DISCUSSIONS

The following sections compare the performance of the OPSS with the PF algorithm for different system capacities and buffering threshold.

5.1 Performance Comparison with Increasing System Capacity

Figure 2 and Figure 3 show the minimum MOS and FDR of the OPSS solution and PF algorithm with increasing system capacity. The buffering threshold of each user was fixed at 200 ms in this performance comparison. It can be observed that the minimum MOS and FDR degrade with increasing system capacity as there are insufficient RBs to schedule the video packets from the users. Table 2 shows that if at least a good perceptual quality has to be satisfied for all users, then the maximum system capacities that the OPSS and PF can simultaneously support are 35 and 25 users respectively. This is equivalent to 40% improvement in the system capacity achieved in the OPSS over the PF algorithm. Moreover, it can be observed in Table 3 that when the system capacity is at 35 users, the OPSS significantly minimizes the FDR by 42.6% compared to the PF algorithm. This indicates that the OPSS solution is superior to the PF algorithm in providing a continuous playback for all video users.

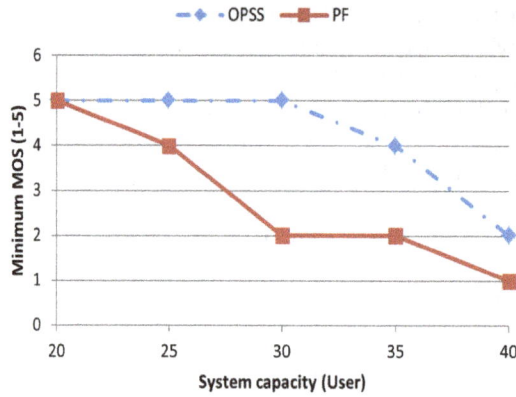

Figure 2. Minimum MOS vs. system capacity

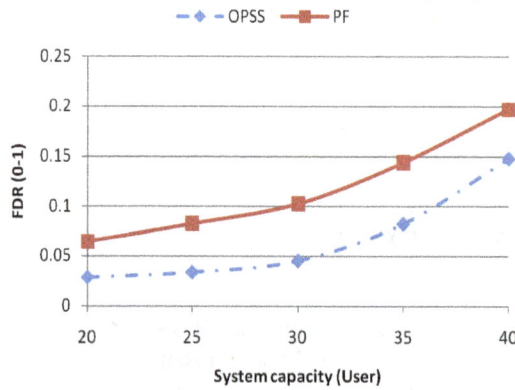

Figure 3. FDR vs. system capacity

Table 2. Maximum system capacities to support a range of minimum MOS

Minimum MOS	Perceptual Quality	Maximum System capacity	
		OPSS	PF
1	Bad	>40	40
2	Poor	40	30
3	Fair	37.5	27.5
4	Good	35	25
5	Excellent	20	20

Table 3. FDR at 35 users

	FDR	Improvement in OPSS over PF (%)
PF	0.143571	42.6
OPSS	0.082446	

The improvements in the minimum MOS and FDR performances in OPSS compared to the PF algorithm can be attributed to the following factors. The buffer management strategy enables the OPSS to efficiently utilize the available RBs by selectively (re)transmitting packets that can be used for decoding and video playback. Furthermore, OPSS integrates the playback buffer information in the packet algorithm so as to ensure a continuous video playback for each user.

5.2 Impact of Buffering Threshold on Performance

The impact of buffering threshold on the OPSS solution and PF algorithm are studied in this section. In this performance comparison, the system capacity was fixed at 25 users (i.e. the system capacity where all users experience at least a good perceptual quality in OPSS and PF – as shown in Figure. 2). As previously discussed in Section 2, the start or resume of a video playback is delayed with increasing buffering threshold because more frames need to be filled in the playback buffer. As there is insufficient time to playback all of the video frames of a user (i.e. the maximum duration of a video session is fixed at T ms for each user), one or more frames of the user may not be able to be played back at the end of its session and this leads to degradations of the minimum MOS and FDR with increasing buffering threshold (as shown in Figure 4 and Figure 5).

Figure 4 shows that the OPSS minimizes degradation due to the impact of the buffering threshold because it is capable of providing an excellent perceptual quality (minimum MOS=5) for all users for up to 600 ms buffering threshold. On the other hand, at least one user in the PF algorithm experienced a bad perceptual quality (minimum MOS=1) at the 600 ms buffering threshold. This implies that the majority of the frames of a number of users cannot be played back when their video sessions end due to the time taken to start or resume the video playback.

When compared with the PF algorithm, it can be observed in Figure 5 that the OPSS significantly minimizes the FDR at a lower buffering threshold. The packet scheduling algorithm of the OPSS gives a higher priority for a user with the least number frames to receive its packets. At a lower buffering threshold, the users in the OPSS can fill their playback buffer with sufficient number of frames earlier than the PF algorithm and hence allowing OPSS to minimize the delay to start or resume the video playback. Table 4 shows that the OPSS minimizes the FDR by 58.7% compared to the PF algorithm at 200 ms buffering threshold.

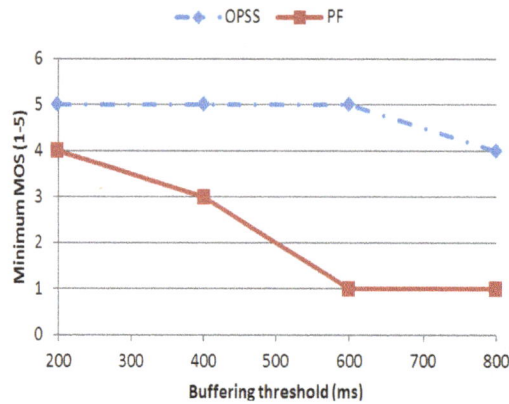

Figure 4. Minimum MOS vs. buffering threshold

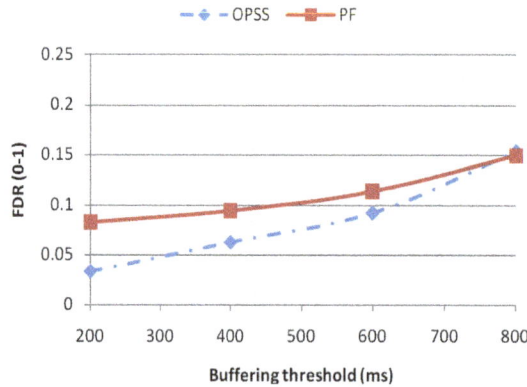

Figure 5. FDR vs. buffering threshold

The PF algorithm is likely to outperform the OPSS in terms minimizing the FDR when the buffering threshold is over 800 ms. This is because there is only a limited number of frames that can be played back in the PF algorithm (i.e. minimum MOS=1 in the PF at 800 ms buffering threshold). Note that the freezing delay is only computed for the frames that can be played back at the users. On the other hand, even if the FDR in the OPSS is likely to be worse than the PF algorithm when the buffering threshold increases above 800 ms, the OPSS guarantees that majority of the frames can be played back throughout each user's session and hence improving the perceptual quality experienced at the users.

Table 4. FDR at 200 ms buffering threshold

	FDR	**Improvement in OPSS over PF (%)**
PF	0.082552	58.7
OPSS	0.034056	

6. CONCLUSION

Simultaneous streaming of multiple video sessions is a challenging task due to the QoS constraint of video streaming as well as unreliable and resource-constrained of the downlink LTE. A number of studies that attempted to address this challenge by improving the perceptual quality or ensuring video playback continuity have been developed. This paper proposes a novel solution that combines the buffer management strategy with packet scheduling algorithm to improve video streaming performance in the downlink LTE. It was demonstrated via computer simulation that the proposed solution is particularly effective in providing a continuous video playback with good perceptual quality for more users. If at least a good perceptual quality is to be satisfied for all users (QoS constraint of video streaming), then the proposed solution improves the system capacity by 40% over the PF algorithm. Moreover, it minimizes the video playback interruption by 42.6% as compared to the PF algorithm. The proposed solution has a low computational complexity and hence suited for implementation in the downlink LTE and other multi-carrier mobile cellular systems without additional hardware cost.

ACKNOWLEDGMENT

This work is supported by International Islamic University Malaysia (IIUM) Endowment Fund Type B (EDW B11-198-0676).

REFERENCES

[1]   Holma, H. & Toskala, A. (2009) LTE for UMTS: OFDMA and SC-FDMA Based Radio Access. John Wiley & Sons Ltd.

[2]   Fan, L., Guizhong, L. & Lijun, H. (2009) Application-Driven Cross-Layer Approaches to Video Transmission over Downlink OFDMA Networks. in IEEE Global Telecommunications Workshops, 1-6.

[3]   Sun, H. Vetro, A. & Xin, J. (2007)An Overview of Scalable Video Streaming. Wireless Communications and Mobile Computing, 7(2), 159-172.

[4]   Seeling, P., Reisslein, M. & Kulapala, B. (2004) Network Performance Evaluation Using Frame Size and Quality Traces of Single-Layer and Two-Layer Video: A Tutorial. IEEE Communications Surveys & Tutorials, 6(3), 58-78.

[5]   Haghani, E. Shyam, P. Doru, C. Eunyoung, K. & Ansari, N. (2009) A Quality-Driven Cross-Layer Solution for MPEG Video Streaming Over WiMAX Networks. IEEE Transactions on Multimedia, 11(6), 1140-1147.

[6]   Ozcelebi, T., Sunay, M.O., Tekalp, A.M. & Civanlar, M.R. (2005) Cross-Layer Design for Real-Time Video Streaming over 1xEV-DO using Multiple Objective Optimization, in IEEE Global Telecommunications Conference, 2761-2766.

[7]   Haghani, E., Ansari, N., Shyam, P. & Doru, C. (2010) Traffic-Aware Video Streaming in Broadband Wireless Networks. in IEEE Wireless Communications and Networking Conference, 1-6.

[8]   Yongjin, C. Kuo, C.C.J., Renxian, H. & Lima, C. (2009) Cross-Layer Design for Wireless Video Streaming. in IEEE Global Telecommunications Conference, 1-5.

[9]   Jalali, A., Padovani, R. & Pankaj, R. (2000) Data Throughput of CDMA-HDR a High Efficiency-High Data Rate Personal Communication Wireless System. in IEEE 51st Vehicular Technology Conference Proceedings, 1854-1858.

[10] Boggia, G., Camarda, P., Fortuna, R. & Grieco, L.A. (2009) A Scheduling Strategy to Avoid Playout Interruptions in Video Streaming Systems. in Proceedings of 18th Internatonal Conference on Computer Communications and Networks. 1-6.

[11] Junhua, T., Liren, Z. & Chee-Kheong, S. (2006) An Opportunistic Scheduling Algorithm for MPEG Video Over Shared Wireless Downlink. in IEEE International Conference on Communications. 872-877.

[12] Honghai, Z., Yanyan, Z., Khojastepour, M.A. & Rangarajan, S. (2009) Scalable Video Streaming over Fading Wireless Channels. in IEEE Wireless Communications and Networking Conference. 1-6.

[13] Tupelly, R.S., Zhang, J. & Chong, E.K.P. (2003) Opportunistic Scheduling for Streaming Video in Wireless Networks. in Conference on Information Sciences and Systems. 1-6.

[14] Seeling, P., Reisslein, M. & Fitzek, F.H.P. (2006) Layered Video Coding Offset Distortion Traces for Trace-Based Evaluation of Video Quality after Network Transport. in 3rd IEEE Consumer Communications and Networking Conference, 292-296.

[15] Zhenzhong, C., Maodong, L. & Yap-Peng, T. (2010) Perception-Aware Multiple Scalable Video Streaming Over WLANs. IEEE Signal Processing Letter. 17(7), 675-678.

[16] Klaue, J., Rathke, B. & Wolisz, A. (2003) Evalvid - A Framework for Video Transmission and Quality Evaluation. in 13th International Conference on Modelling Techniques and Tools for Performance Evaluation. 1-5.

[17] Guanfeng, L. & Ben, L. (2008) Effect of Delay and Buffering on Jitter-Free Streaming Over Random VBR Channels. IEEE Transactions on Multimedia, 10(6), 1128-1141.

[18] Vukadinovic, V. & Karlsson, G. (2010) Video Streaming Performance under Proportional Fair Scheduling. IEEE Journal on Selected Areas in Communications, 28(3),399-408.

[19] Ozcelebi, T., Oguz, S.M., Murat, T.A. & Reha, C.M. (2007) Cross-Layer Optimized Rate Adaptation and Scheduling for Multiple-User Wireless Video Streaming. IEEE Journal on Selected Areas in Communications, 25(4), 760-769.

[20] Larmo, A., Lindstrom, M., Meyer, M., Pelletier, G., Torsner, J. & Wiemann, H. (2009) LTE Link-Layer Design. IEEE Communications Magazine, 47(4), 52-59.

[21] Andrews, M., Kumaran, K., Ramanan, K., Stolyar, A., Whiting, P. & Vijayakumar, R. (2001) Providing Quality of Service over a Shared Wireless Link. IEEE Communications Magazine, 39(2), 150-154.

[22] Tsybakov, B.S. (2002) File Transmission over Wireless Fast Fading Downlink. IEEE Transactions on Information Theory, 48(8), 2323-2337.

[23] Ramli, H. A. M. & Sandrasegaran, K. (2013) Robust Scheduling Algorithm for Guaranteed Bit Rate Services. International Journal of Mobile Communications, 11(1), 71-88.

[24] Pokhariyal, A., Monghal, G., Pedersen, K.i., Mogensen, P.E., Kovacs, I.Z., Rosa, C. & Kolding, T.E. (2007) Frequency Domain Packet Scheduling Under Fractional Load for the UTRAN LTE Downlink. in IEEE Vehicular Technology Conference, 699-703.

[25] Rappaport, T.S. (2003) Wireless Communications: Principles and Practice. Prentice Hall.

[26] Gudmundson, M. (1991) Correlation Model for Shadow Fading in Mobile Radio Systems. Electronics Letters, 2145-2146.

[27] Komninakis, C. (2003) A Fast and Accurate Rayleigh Fading Simulator. in IEEE Globecom, 3306-3310.

[28] Pedersen, K.I., Monghal, G., Kovacs, I.Z., Kolding, T.E., Pokhariyal, A., Frederiksen, F. & Mogensen, P. (2007) Frequency Domain Scheduling for OFDMA with Limited and Noisy Channel Feedback. in Vehicular Technology Conference, 1792-1796.

[29] Ramli, H.A.M. Sandrasegaran, K., Basukala, R., Patachaianand, R. & Afrin, T.S (2011) Video Streaming Performance under Well-Known Packet Scheduling Algorithms. International Journal of Wireless & Mobile Networks (IJWMN), 3(1), 25-38.

[30] Ruangchaijatupon, N. & Ji, Y. (2008) Simple Proportional Fairness Scheduling for OFDMA Frame-Based Wireless Systems. in IEEE Wireless Communications and Networking Conference, 1593-1597.

[31] Mugen, P. & Wenbo, W. (2004) Advanced HARQ and Scheduler Schemes in TDD-CDMA HSDPA Systems. in Joint Conference of the 10th Asia-Pacific Conference on Communications and the 5th International Symposium on Multi-Dimensional Mobile Communications, 67-70.

[32] Wang, Y. & Yang, H. (2003) Retransmission Priority Scheduling Algorithm for Forward Link Packet Data Service. in International Conference on Communication Technology Proceedings, 926-930.

[33] Ramli, H.A.M., Basukala, R., Sandrasegaran, K. & Patachaianand, R. (2009) Performance of Well Known Packet Scheduling Algorithms in the Downlink 3GPP LTE System. in IEEE 9th Malaysia International Conference on Communications (MICC), 815-820.

13

OPTIMIZED COOPERATIVE SPECTRUM-SENSING IN CLUSTERED COGNITIVE RADIO NETWORKS

Birsen Sirkeci-Mergen and Wafa-Iqbal

Electrical Engineering, San Jose State University, San Jose, CA

birsen.sirkeci@sjsu.edu, wafa.iqbal1@gmail.com

ABSTRACT

In cognitive radio networks, the pertinent task of spectrum sensing at the Secondary Users (SUs) can be achieved when the SUs cooperate in order to make a final decision about the presence of a communicating Primary User (PU). In this paper, we study a two-hop relaying system in which SUs are grouped into D clusters. The SUs transmit a simple power function (parameterized by p) of their observationto a Fusion Centre (FC) using D orthogonal channels. The FC combines the receptions from cooperating nodes linearly. The goal of this work is to maximize the probability of detection over the parameters D (number of clusters), p (power function exponent), and w(linear combining coefficients) for a given false alarm probability. Overall, this work quantifies the advantages of optimal cooperation in primary detection in cognitive radio networks.

KEYWORDS

Wireless Networks, Relaying, Spectrum Sensing, Cognitive Radios, Cooperation, Fusion Centre.

1. INTRODUCTION

Wireless communication is progressing at an accelerated speed. Increasing variety of applications and features of wireless devices is leading to demands for higher and higher data rates. However, the bandwidth licensed to radio communication is limited. The infamous question is "How do we get better data rates under limited bandwidth requirements to meet the demand?".Efficient spectrum utilization is the key to answer this question.

In 2002, Federal Communications Commission (FCC), the US government agency that regulates the use of frequency bands of the electromagnetic spectrum, indicated that the licensed frequency bands are unused 90% of the time[1]. In 2008, FCC ruled that unused portions of the RF spectrum will be made available for public use under certain conditions. In the light of this rule, spectrum efficiency can be improved if radio devices are equipped with technologies that take advantage of the licensed spectrum when it is unused. An emerging advanced solution for efficient spectrum utilization is the so-called cognitive radios.

A cognitive radio (CR) is a transceiver technology in which frequency spectrum is continuously sensed for unoccupied spaces. In a CR system, the primary user (PU) is the one who has licensed privilege to transmit in a particular frequency band and other users known as secondary users (SU) are the unlicensed users who desire to share the spectrum. The available unused frequency bands are called 'spectrum holes'. SUs sense the spectrum for spectrum holes continuously. A CR is capable of not only sensing the spectrum, but also, monitoring, detecting and adapting its communication channel access. For example, a CRcan intelligently adjust its transmission parameters according to the availability in the frequency bands[2],[3]. CRtechnology has gained a lot of attention in the last decade. Currently, communication standards are adaptingthis technology [4].

Cooperative spectrum sensing is a scheme in which SUs cooperate with each other in a distributed or centralized manner, in order to make the decision about spectrum availability. This could be done via a Fusion Centre (FC). The SUs sense the channel for the presence of PUs and relay a function of their observations to the FC for a collective decision. The choice of this relaying is critical in order to optimize the overall performance at the FC.In the next subsection, we summarize the recent relevant work on cooperative spectrum sensing.

1.1. Cooperative Spectrum Sensing

In cooperative spectrum sensing, the SUs collaborate with each other in sensing the spectrum [5]. If optimized, cooperation reduces the power requirements at the SUsand improves the sensing performance even if it may introduce overhead for certain cases. In the case when SUs cooperate through a FC, every SU transmits its received signal to the FC that makes a decision about the presence of a PU based on the collective information from all the SUs.This is also called relay-assisted cooperative spectrum sensing [6].

The transmissions of the SUs to the fusion center could be on orthogonal channels [6], [7]. In this case, each SUforwards a function of their observation to the fusion center through and individualorthogonal channel similar to the well-known time-division multiple-access (TDMA), or frequency-division multiple-access (FDMA). On the other hand, transmissions of the SUs to the fusion center could be non-orthogonal, that is cooperating SUs transmit a function of their observation byusing the same channel. In the non-orthogonal channel model, it is assumed that SUs are synchronized so that the received signal in the fusion center is the coherent sumof transmitted signals by SUs[8], [9]. For orthogonal channels, the fusion center canuse various combining techniques of the received vector to obtain the final decision. It is shown in [10][11] that the probability of error for coherent orthogonal channel system will not improve with the increasing number of SUs. On the otherside, the performance of the non-orthogonal channel systemimproves with the increasing number of SUs due to the array gain [9], [10].

It iswell-known that in order to have an energy-efficient and reliable spectrum sensing, it is important for SUs to cooperate with each other when sensing for the PUs. However, one has to carefully weigh the trade-offs between the achievable *Cooperative Gain* and the incurred *Cooperative Overhead*[12]. In the case of orthogonal access between SUs and a FC, each radio is dedicated anorthogonal channel, and the requirement for bandwidth scales by the number of SUs. Then, the receptions from SUs at the FC are combined. In general, the linear combining techniquesare attractive, because, they are simple compared to non-linear techniques, and when the weighting coefficients are optimized, the improvement in the probability of detection at the fusion center is significant. Onthe other hand, in the case when non-orthogonal access is utilized from SUs to the FC, bandwidth requirements are negligible. Furthermore, the additive noise in the non-orthogonal channelis negligible, especially for large networks, compared to orthogonal channels since it is independent of the number of SUs. The gains due to optimized weighting coefficients in orthogonal channels and the independence of noises from the number of SUs in non-orthogonal channels posea trade-off. In order to optimizethis trade-off, one scheme proposed in [13]by the first author: group-orthogonal multiple access channel (MAC) approach for spectrum sensing. In group- orthogonal MAC, SUs utilize the available orthogonal channels in clusters, and each SU transmit to the FC the energy of its reception from the PUs. In [13], authors exploit the benefits of both orthogonal and non-orthogonal transmissions byfinding the optimal number of users that should be in an orthogonal group and the optimal linear weighting coefficients at the FC.

In this paper, we study optimal relaying function at SUs under different channel access schemes from the SUs to the FC. The considered cases are orthogonal, non-orthogonal and group-orthogonal multiple-access channels (MACs). In the group-orthogonal case, SUs are clustered into D groups that transmit on D orthogonal channels. In fact, orthogonal MAC and non-orthogonal MAC are special cases of group-orthogonal MAC when D=number of SUs, and D=1, respectively. The expressionsfor the probability of detection as a function of probability of false alarm under different group sizes and mappings arederived and analysed. This paper optimizes the performance over a set of relay functions and channel access schemes. This will help the SUs to make intelligent decisions when selectinghow and what to send to FC,for given a probability of false alarm in detecting the spectrum availability.

The rest of the paper is organized as follows. In Section 2, we give the problem formulation. In Section 3, we derive theoptimal number of groups and weighting coefficientunder certain assumptions. Simulationresults are given in Section 4. Finally, Section 5 concludesthe paper.

2. SYSTEM MODEL

We consider a cognitive radio network that is composed of aPU, multiple SUs and a FC-which could also be one of the SUs (see Fig. 1). Although the spectrum band under consideration is licensed to the PUs, they may or may not be transmitting during the considered time-slot. Hence, SUs need to decide whether thePUis idle (null hypothesis) or it is using the channel (alternative hypothesis) in order to utilize the band efficiently. In the considered set-up, the decisions are made cooperatively- that is each user makes decisions based on receptions from multiple SUs which also serve as relays. When acting as relays, each SU makes an observation, and transmits a signal based on solely its observation to the FC. We assume the FC combines the received signals linearly and makes final decision about the existence of the primary based on the combined signal. Linear combining at the FC is an attractive method primarily due to its simplicity. The two hypotheses: H_0 (no primary user exists) andH_1 (at least one primary user exist) form a binary hypothesis test given as below:

$$H_0: x_i(k) = v_i(k) \qquad i=1,....M, \quad k=1,....N$$
$$H_1: x_i(k) = h_i s(k) + v_i(k) \qquad i=1,....M, \quad k=1,....N$$

where$x_i(k)$ is the observed signal by the ith secondary user over N timeslots, $s(k)$ is the transmitted signal by the PU in the kth timeslot and $v_i(k)$is the additive noise at the ith user in the kth timeslot. The noise$v_i(k)$ is assumed be white Gaussian noise with zero mean and variance $\sigma^2$and also $v_i(k)$sare assumed to be independent and identically distributed (i.i.d.) over time index k and user index i. The channel gains from the PU to the SUsare assumed to stay constant over the observation interval (slow fading scenario).

In the network, each SU observes the channel for N timeslots and then forwards a power function of the observed signal to the FC: $u_i = f(x_i) = \beta_i |x_i|^p$ where $\beta_i = P_i / \sqrt{E\{|x_i|^2\}}$ is the scaling factor so that average transmission power is boundedbythe power constraint P_i. A common relay operation is to send the energy of the observedsignal [13], which is equivalent to the case when p

is chosen to be 2, and β_i =1 in our scenario. Our goal is to optimize over the power function exponent p so that the performance is improved.

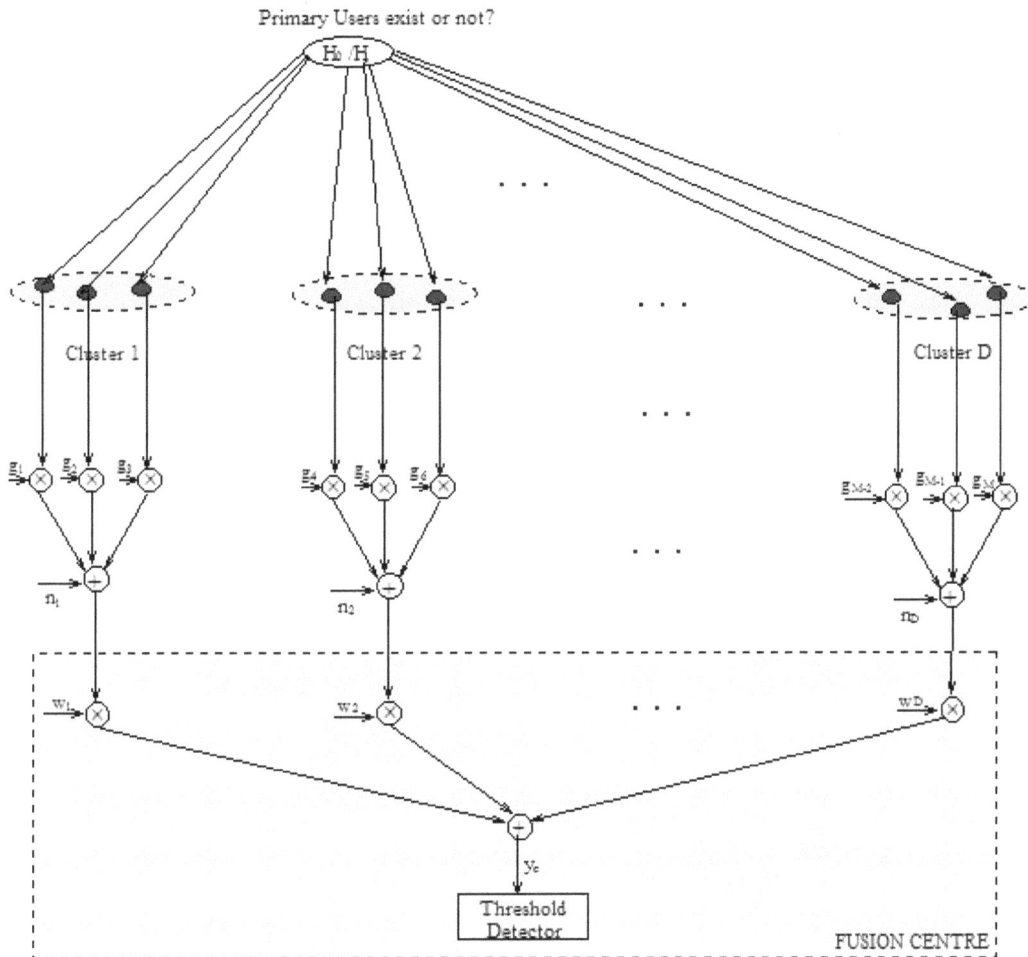

Figure 1. Group-orthogonal MAC with M = 3D secondary users and D clusters

In the network, SUsare grouped into Dclusters (see Fig. 1). Each cluster is dedicated to on an orthogonal channel, and users in the same cluster transmit on the same orthogonal channel. We call this set-up *group-orthogonal MAC (Multiple Access Channel)*. The clusters are assumed to be pre-determined. For example, one could form clusters based on geographical proximity or signal quality. However, the question of how the clusters are formed is out of scope of this paper. Let S_j denote a set of users in the jth group where $j=1...D$. For simplicity in the analysis, we also assume that clusters have equal number of nodes. Then, the combined signal in the jth orthogonal channel can be written as:

$$y_j = \sum_{m \in S_j} g_m u_m + n_j \quad j = 1,....D$$

where g_m is the channel gain from SU to the FC and n_j is the noise added at each channel and is assumed to be i.i.d. white Gaussian noise with zero mean and variance σ^2. When information from each group reaches the FC, it is linearly combined after being weighted. The weighting vector is defined as $\mathbf{w} = [w_1, w_2, \ldots, w_D]$.

After combining, the signal observed at the FC is denoted by:

$$y_c = \sum_{j=1}^{D} w_j y_j = \sum_{j=1}^{D} w_j \left(\sum_{m \in S_j} g_m u_m \right) + \sum_{\neq 1}^{D} n_j w_j$$

At the FC, the global test statistic y_c is compared with γ_c to make a decision about PUs, that is

if $y_c \geq \gamma_c$, decide that H_1 has occurred,

if $y_c < \gamma_c$, decide that H_0 has occurred.

3. OPTIMIZED COOPERATIVE SPECTRUM-SENSING

In this section, first we describe the performance metrics that are used, and then we describe the optimization problem. Solution for the optimization problem is also provided.

3.1. Performance Metrics

We use the two important metrics: probability of detection $P_d = P(H_1 | H_1)$ and probability of false alarm $P_f = P(H_1 | H_0)$. Our goal is to maximize the probability of detection, P_d, for a given probability of false alarm, P_f. The optimization is over the set of parameter: as the number of orthogonal channels D, the relaying function $\beta_i |x_i|^p$, and weighting coefficients w_js.

In the following, we will use central limit theory [14] to derive analytical expressions for P_d and P_f. We can argue that for large N (the observation time interval), x_i can be assumed to be asymptotically normally distributed as well as y_js and y_c. For a normally distributed random variable y_c, the probability of detection and false alarm can be expressed as follows for a given detection threshold γ_c at the FC:

$$P_d = P[H_1 | H_1] = P[y_c \geq \gamma_c | H_1] = Q\left[\frac{\gamma_c - E[y_c | H_1]}{\sqrt{Var[y_c | H_1]}} \right] \tag{1}$$

$$P_f = P[H_1 | H_0] = P[y_c \geq \gamma_c | H_0] = Q\left[\frac{\gamma_c - E[y_c | H_0]}{\sqrt{Var[y_c | H_0]}} \right] \tag{2}$$

where $Q(x) = 1/\sqrt{2\pi} \int_x^\infty e^{-u^2/2} du$ denotes the Q-function.

In order to find detection and false alarm probabilities based on the above mentioned formulas it is required to find the conditional mean and variances under both hypotheses for a given parameter set. Note that the P_d is actually a function of not only P_f, but also a function network parameters such as the number of cluster (D), the relay function exponent (p), the FC combining coefficients (w), channel coefficients between primary and cognitive radios (h_is), the channel coefficients between the cognitive radios and the FC (g_is), transmission power of the radios (P_is), and the noise powers σ^2 and δ^2. Our goal is to maximize the probability of detection over the parameters: (i) the number of clusters (D); $(ii)$the relaying function exponent (p); and (iii) weighting coefficients at the FC, w, when the rest of the parameters are given.This implies that the channel state information (CSI) is known for both links between PUs and SUs, and SUs and the FC. Note that the channel fading coefficients h_is and g_is are assumed to be slowly varying; hence the CSI assumption is sensible. The following lemma provides an explicit analytical expression for P_d as a function of P_f.

Lemma 1: For given probability of false alarm P_f, number of clusters D, relaying function exponent p, and weighting coefficients w, if the channel gains $|h_i|^2$s are all equal ($|h_i|^2 = \alpha$, $\forall i$), then the probability of detection P_d is given as follows for large N:

$$P_d(D,p,w) = Q\left[\frac{E\left[y_c|H_0\right]-E\left[y_c|H_1\right]+Q^{-1}\left(P_f\right)\sqrt{Var\left[y_c|H_0\right]}}{\sqrt{Var\left[y_c|H_1\right]}}\right] \tag{3}$$

where

$$E[y_c \mid H_0] = A_p g_D^H w, \quad E[y_c \mid H_1] = B_p g_D^H w,$$

$$Var[y_c \mid H_0] = C_p w^H G_D w + \delta^2 w^H w,$$

$$Var[y_c \mid H_1] = D_p w^H G_D w + \delta^2 w^H w,$$

and

$$g_D = [\sum_{i\in S_1} g_i\sqrt{P_i},...,\sum_{i\in S_D} g_i\sqrt{P_i}], \quad G_D = \mathrm{diag}(\sum_{i\in S_1} g_i^2 P_i,...,\sum_{i\in S_D} g_i^2 P_i).$$

The coefficients A_p, B_p, C_p, and D_p depend on the relay function exponent p and are given as follows:

$$A_p = \Gamma\left(\frac{p+1}{2}\right)\Bigg/ \sqrt{\frac{\sqrt{\pi}}{N}\Gamma\left(\frac{2p+1}{2}\right)+\frac{N-1}{N}\Gamma^2\left(\frac{p+1}{2}\right)}$$

$$B_p = \left(\sum_{k=0}^{N-1}\Phi_{p,k}\right)\Bigg/\sqrt{\pi^{1/2}\frac{\Gamma\left(\dfrac{2p+1}{2}\right)}{\Gamma^2\left(\dfrac{p+1}{2}\right)}\left(\sum_{k=0}^{N-1}\Phi_{2p,k}\right)+2\sum_{k_1<k_2}\Phi_{p,k_1}\Phi_{p,k_2}}$$

$$C_p = \left(\Gamma\left(\frac{2p+1}{2}\right)-\frac{1}{\sqrt{\pi}}\Gamma^2\left(\frac{p+1}{2}\right)\right)\Bigg/\left(\Gamma\left(\frac{2p+1}{2}\right)+\frac{N-1}{\sqrt{\pi}}\Gamma^2\left(\frac{p+1}{2}\right)\right)$$

$$D_p = \left(\Gamma\left(\frac{2p+1}{2}\right)\sum_{k=0}^{N-1}\Phi_{2p,k}-\frac{1}{\sqrt{\pi}}\Gamma^2\left(\frac{p+1}{2}\right)\sum_{k=0}^{N-1}\Phi_{p,k}^2\right)\Bigg/\left(\Gamma\left(\frac{2p+1}{2}\right)\sum_k\Phi_{2p,k}+\frac{2}{\sqrt{\pi}}\Gamma^2\left(\frac{p+1}{2}\right)\sum_{k_1<k_2}\Phi_{p,k_1}\Phi_{p,k_2}\right)$$

where $\Phi_{p,k} = {}_1F_1\left(\dfrac{-p}{2},\dfrac{1}{2},\dfrac{-\alpha\, s^2(k)}{2\sigma^2}\right)$, $\Gamma(x)$ denotes the gamma function, and ${}_1F_1$ denotes the confluent hyper-geometric function[15].

Proof: Using Eqn. (2), for any givenP_f, threshold $\gamma_c$can be written as:

$$\gamma_c = E\left[y_c|H_0\right]+Q^{-1}\left(P_f\right)\sqrt{Var\left[y_c|H_0\right]} \tag{4}$$

By substituting Eqn. (4) in Eqn. (1), we obtain Eqn. (3). The derivations for $E[y_c/H_0]$, $E[y_c/H_1]$, $Var[y_c/H_0]$, and $Var[y_c/H_1]$ are given in the Appendix 3.1. ◆

It is important to note that Lemma 1 provides a formulation in which the dependences on the optimization parameters (D, p, w) are partially decoupled. The variables w, g_D and G_D depend on the cluster size D, the constants A_p, B_p, C_p, and D_p depend on the relay function exponent p, and the weighting coefficient w shows up explicitly in the expression. This will help solve the optimization problem.

3.2. Optimized Cooperative Transmission and Reception

We formulate the optimization problem as follows. Given the channel gains ($|h_i|^2$s, and g_is), the transmission powers of the SUs (P_is), and the noise powers σ^2 and δ^2 the goal is to maximize the P_d for a given limit on false alarm probability P_f:

$$\max_{D,p,w} P_d(D,p,w)$$

We make the following assumptions in order to solve this problem.

1. Uniform channel gains: $|h_i|^2 = \alpha$, and $g_i = \beta > 0$.
2. Uniform transmission powers: $P_i = P$ for all i.
3. The orthogonal groups have equal number of SUs (assuming M/D is an integer).
4. The weighting coefficients, w_is, are nonnegative.

Under these assumptions the following theorem provides the optimal D, and w for a given p value.

Theorem 1: In the group-orthogonal MAC system, for a given relay power function with exponent p, if channel gains are equal ($|h_i|^2 = \alpha, g_i = \beta$), then the optimal weighting coefficients that maximizes P_d (Eqn. (3)) are uniform for a given D, that is

$$w_i = 1/D, i = 1 \ldots D.$$

And the optimal D for given w and p is given as

$$D \cong \begin{cases} 1 & P_f > Q\left(\dfrac{(A_p - B_p)\sqrt{C_p \beta^2 PM + \delta^2}}{(D_p - C_p)\beta\sqrt{P}}\right) \\[3em] M & P_f < Q\left(\dfrac{(A_p - B_p)\sqrt{C_p \beta^2 PM + \delta^2 M}}{(D_p - C_p)\beta\sqrt{P}}\right) \\[3em] \left.\dfrac{Q^{-2}(f)(D_p - C_p)^2 \beta^2 P}{(A_p - B_p)^2 \delta^2} - \dfrac{C_p \beta^2 PM}{\delta^2}\right|_M & \text{otherwise} \end{cases}$$

where $\underline{X}\big|_M$ denotes the divisor of M that is closed to X.

Proof: See Appendix 3.2. ◆

The theorem states optimal linear combining coefficients at the FC should be uniform for D orthogonal channels and should sum to 1. This is very intuitive due to the assumptions on the equal channel gains and noise powers, and also identical relay functions at the relays. On the other hand, optimal D has three different regions: (i) when the false alarm probability is high $(P_f > P_L)$, the optimal D is equal to 1 which implies that non-orthogonal transmission is optimal; (ii) when the false alarm probability is low $(P_f < P_H)$, the optimal D is equal to M which implies that each SU should transmit on an orthogonal channel and no clustering of SUs; and (iii) when the false alarm probability is between P_L and P_H, optimal scheme is group-orthogonal transmission. Note that for some special scenarios, the third region may merge with one of the other regions, that is the rounding operation $\rfloor_M$ in the above equation may lead to $D=1$ or $D=M$.

Optimization over the relaying function exponent p can be done by replacing the optimal values for D and $\mathbf{w}$ obtained in Theorem 1 in Eqn. 3, and by using an optimization toolbox for nonlinear integer programming.

4. SIMULATIONS

In this section, we provide probability of detection versus probability of false alarm curves for different values of p and D. For all the simulations, the number of users M=4, observation time N=1, the channel gains $|h_i|^2 = \alpha = 100$, $g_i = \beta = 1$, relay transmission powers P =1, and noise powers $\sigma^2 = 1$, and $\delta^2 = 5$. Below, P_d denotes the probability of detection and P_f denotes the probability of false alarm. We assume the primary signal $|s(k)|^2 = 1/N$, for all k, for simplicity.

In Fig.2, we display the P_d as a function of P_f when the relay function has exponent $p=1$ and $p=3$. The curves for various channel access scenarios between relays and FC are shown: orthogonal access ($D=M=4$), non-orthogonal access ($D=1$), and group-orthogonal access ($D=2$). It is observed that for lower probability of false alarms, orthogonal MAC gives the best probability of detection, and for higher probability of false alarms, non-orthogonal MAC gives the best probability of detection. Using Theorem 1, we can obtain the boundaries of these two different regions: $P_f > 0.2797$ and $P_f < 0.2180$ for $p=1$ and $P_f > 0.1180$ and $P_f < 00885$ for $p=3$, which are consistent with the simulations. In Fig. 3, we display the zoomed curves corresponding to the region $0.2180 < P_f < 0.2797$ for $p=1$. According to Theorem 3, in this region optimal D could be 1, 2, and 4 which is what we observe in Fig. 3. Overall, the relay function with exponent p=3 outperforms the relay function with exponent p=1. Furthermore, we observe that the range of P_f where group-orthogonal MAC is optimal is getting smaller with the increase in p.

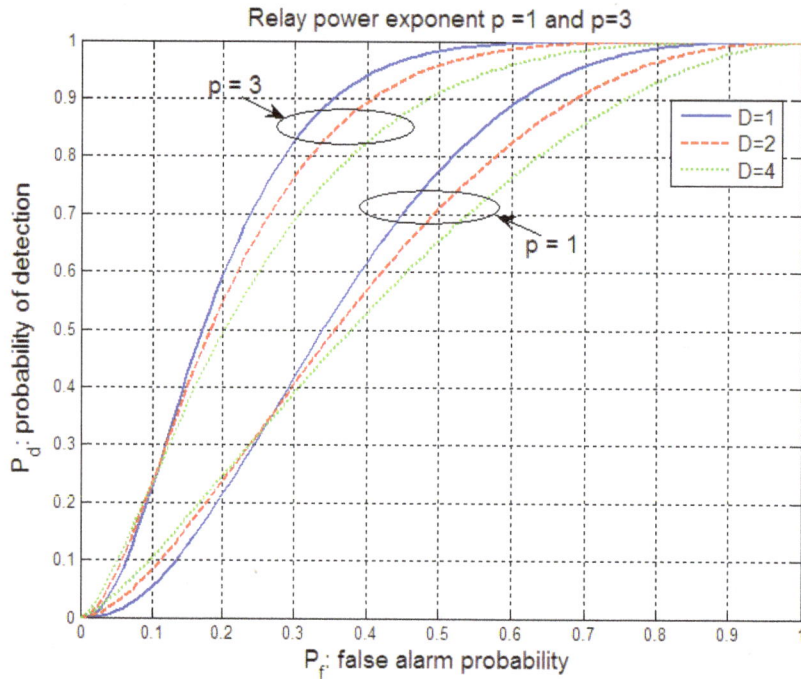

Figure 2 Probability of detection (P_d) vs. Probability of false alarm (P_f) with p =1 and p=3

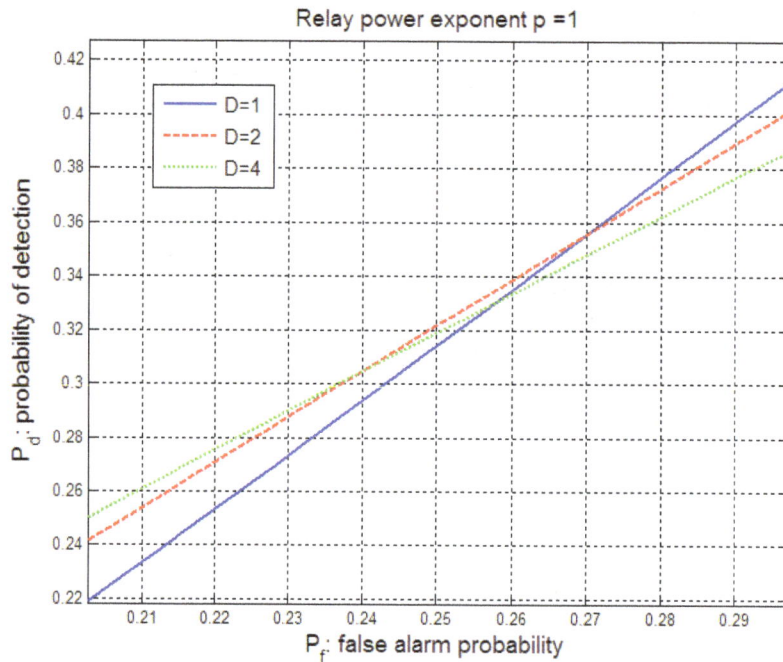

Figure 3 $P_d$vs.P_f for different ranges of P_f: (0.21<P_f<0.29)

Next, we analyse different relay functions for a given channel access scheme in detail. In Fig. 4 and Fig.5, we plot the P_d vs.P_f curves for the non-orthogonal MAC (D=1), and orthogonal MAC(D=M=4), respectively.It can be concluded that for a given D, there does not exist a single relay function that performs optimally for all P_f values. In the limit where relay function exponent p is large, the curves reduces to $P_d = P_f$ line for any D value. Similar behaviour is observed for other D value.

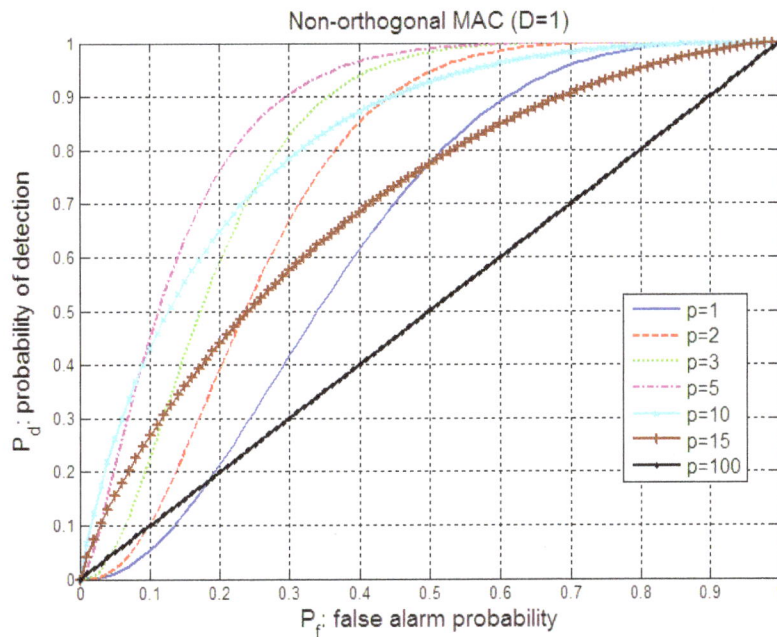

Figure 4 P_d vs. P_f for non-orthogonal channel (D = 1)

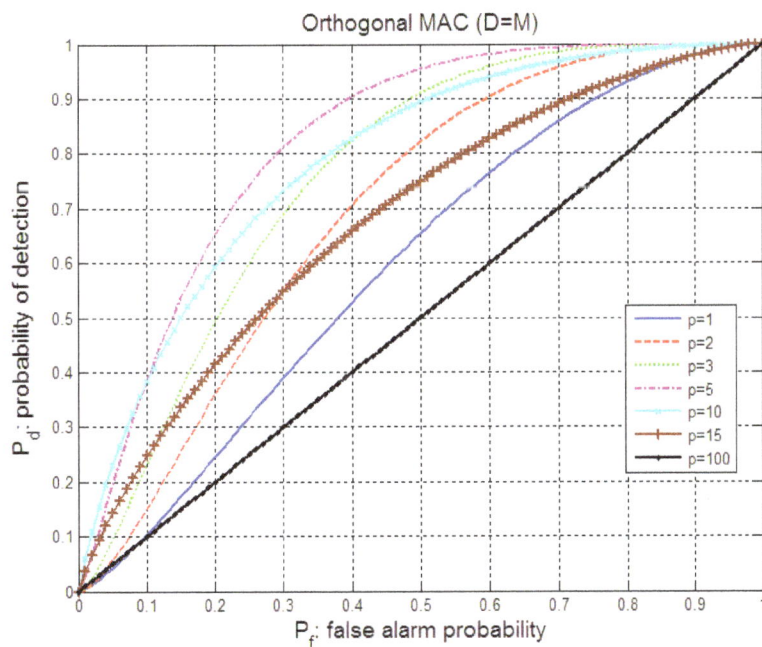

Figure 5 Pd vs. Pf for orthogonal channel (D = M)

In Table 1, we display the optimal relay function exponent and optimal cluster size D for various P_f values. It is important to note that optimal D is always equal to 1, which implies that when optimal relay function (or equivalently p) is selected for a given false alarm probability, then the non-orthogonal scheme becomes optimal globally. In addition, as P_f increases, the optimal relay function exponent p decreases.

Table 1 Optimal D and p

Range of P_f	Optimal p	Optimal D
$0 < P_f < 0.001$	>11	1
$0.001 < P_f < 0.003$	11	1
$0.004 < P_f < 0.009$	10	1
$0.010 < P_f < 0.026$	9	1
$0.027 < P_f < 0.059$	8	1
$0.060 < P_f < 0.126$	7	1
$0.127 < P_f < 0.250$	6	1
$0.251 < P_f < 0.487$	5	1
$0.488 < P_f < 0.897$	4	1
$0.898 < P_f < 0.998$	3	1
$0.999 < P_f < 1.000$	2	1
$P_f = 1$	1	1

Overall, the optimal relay function for any of the channel access schemes is always the power function with high exponents for lower probability of false alarms. However, if the system is robust enough to handle higher probability of false alarms, small power exponents such as $p = 1, 2$ or 3 is the optimal choice of for the relay functions. It can also be concluded that globally non-orthogonal scheme is optimal under the given assumptions.

5. CONCLUSIONS

In this paper, we studied a cognitive radio network in which SUs cooperate in order to make a decision about the primary existence. The proposed scheme is distributed in the sense that the cooperating SUs transmit a power function (parameterized with exponent p) of their local observation, hence does not require any overhead due to cooperation. The SUs transmit to a FC (which could also be one of the SUs) over D orthogonal channels, and FC combines these receptions linearly using weighting coefficients w. We provided analytical solutions and simulations for maximizing the probability of detection at the fusion centre for a given false alarm probability over the parameters D, p, and w. It is interesting that non-orthogonal channel access becomes optimal globally when the best relaying function is utilized even though the orthogonal or group-orthogonal access schemes require more bandwidth. This behaviour is not observed in cooperation strategies where relays simply send their energy to the fusion centre [13]. In summary, this work shows the importance of optimization in cooperative cognitive radio networks in order to extract the gains of cooperation for spectrum sensing with negligible overhead.

6. APPENDICES

6.1. Proof of Lemma 1

Derivation of $E\left[y_c | H_0\right]$: For the first hypothesis H_0 we derive the expected value as:

$$E\left[\sum_{k=0}^{N-1} |x_i(k)|^p | H_0\right] = E\left[\sum_{k=0}^{N-1} |v_i(k)|^p\right] = \sum_{k=0}^{N-1} E\left[|v_i(k)|^p\right]$$

We know that $v_i(k)$ is a zero-mean Gaussian $N(0,\sigma_i^2)$ and for such a random variable the expected value of the absolute function is given by [16]:

$$E\left[\left|\mathbf{x}\right|^p\right] = \frac{1}{\sqrt{\pi}} \; \sigma^p . 2^{\frac{p}{2}} . \Gamma\left(\frac{p+1}{2}\right)$$

Hence,

$$\sum_{k=0}^{N-1} E\left[\left|v_i(k)\right|^p\right] = \sum_{k=0}^{N-1} \frac{\sigma^p 2^{\frac{p}{2}} \Gamma\left(\frac{p+1}{2}\right)}{\sqrt{\pi}} = \frac{N}{\sqrt{\pi}} \; \sigma^p 2^{\frac{p}{2}} \Gamma\left(\frac{p+1}{2}\right) \quad (5)$$

Derivation of Normalization factor for Hypothesis H_0: All the above derived expressions are yet not normalized there we need to find expressions for the normalization factors for every hypothesis. The normalization factor hypothesis H_0 can be derived as follows:

$$\sqrt{E\left[\left(\sum_{k=0}^{N-1}\left|x_i(k)\right|^p\right)^2 \middle| H_0\right]} = \sqrt{E\left[\left(\sum_{k=0}^{N-1}\left|v_i(k)\right|^p\right)^2\right]}$$

It can be easily derived that the expected value of square of sums is given by:

$$E\left[\left(\sum_{k=0}^{N-1}\left|X_k\right|^p\right)^2\right] = \sum_{k=0}^{N-1} E\left[\left|X_k\right|^{2p}\right] + \sum_{k_1<k_2} E\left[\left|X_{k_1}\right|^p\right] E\left[\left|X_{k_2}\right|^p\right] \quad (6)$$

Since $v_i(k)$ is independent and identically distributed for each k, substituting the values of the expectations in the above equation gives:

$$E\left[\left(\sum_{k=0}^{N-1}\left|x_i(k)\right|^p\right)^2 \middle| H_0\right] = \frac{N}{\sqrt{\pi}}\sigma^{2p}2^p\Gamma\left(\frac{2p+1}{2}\right) + \frac{N(N-1)}{\pi}\sigma^{2p}2^p\Gamma^2\left(\frac{p+1}{2}\right) \quad (7)$$

Since y_c is a linear combination of x_i after being scaled by channel gain and weighting factor, so the expected value of y_c can be written as:

$$E\left[y_c \middle| H_0\right] = \frac{\dfrac{N}{\sqrt{\pi}} \; \sigma^p 2^{\frac{p}{2}}\Gamma\left(\frac{p+1}{2}\right)}{\sqrt{\dfrac{N}{\sqrt{\pi}}\sigma^{2p}2^p\Gamma\left(\frac{2p+1}{2}\right) + \dfrac{N(N-1)}{\pi}\sigma^{2p}2^p\Gamma^2\left(\frac{p+1}{2}\right)}} \; \mathbf{g}^H\mathbf{w} \quad (8)$$

Derivation of $E\left[y_c \middle| H_1\right]$: For the second Hypothesis, H_1 we derive the expected value as

$$E\left[\sum_{k=0}^{N-1}\left|x_i(k)\right|^p \middle| H_1\right] = \sum_{k=0}^{N-1} E\left[\left|h_i s(k)+v_i(k)\right|^p\right]$$

We see that $x_i(k)$ is a non-zero mean Gaussian random variable $N(h_i s(k), \sigma_i^2)$ and for a Gaussian random variable with mean μ_x and variance σ_x^2

$$E\left[\left|X\right|^p\right] = \frac{\sigma_x^p . 2^{\frac{p}{2}} . \Gamma\left(\frac{p+1}{2}\right)}{\sqrt{\pi}} ._1F_1\left(\frac{-p}{2},\frac{1}{2},\frac{-1}{2}\left(\frac{\mu_x}{\sigma_x}\right)^2\right)$$

where $_1F_1$ is the confluent hyper geometric function. Hence,

$$\sum_{k=0}^{N-1} E\left[\left|h_i s(k) + v_i(k)\right|^p\right] \quad = \sum_{k=0}^{N-1} \frac{\sigma^p . 2^{\frac{p}{2}} . \Gamma\left(\frac{p+1}{2}\right)}{\sqrt{\pi}} {}_1F_1\left(\frac{-p}{2}, \frac{1}{2}, \frac{-1}{2} \frac{|h_i|^2 . |s(k)|^2}{\sigma^2}\right)$$

Using the assumptions we have made about the channel gain h_i, we can simplify the expression as following:

$$\sum_{k=0}^{N-1} E\left[\left|h_i s(k) + v_i(k)\right|^p\right] = \frac{1}{\sqrt{\pi}} \sigma^p 2^{\frac{p}{2}} \Gamma\left(\frac{p+1}{2}\right) \sum_{k=0}^{N-1} {}_1F_1\left(\frac{-p}{2}, \frac{1}{2}, \frac{-\alpha \, s^2(k)}{2\sigma^2}\right)$$

Derivation of Normalization factor for Hypothesis H_1: Using(6), the normalization factor hypothesis H_1 can be derived as follows:

$$E\left[\left(\sum_{k=0}^{N-1} |x_i(k)|^p |H_1\right)^2\right] = \sum_{k} \frac{\sigma^{2p} 2^{\frac{2p}{2}} \Gamma\left(\frac{2p+1}{2}\right)}{\sqrt{\pi}} {}_1F_1\left(\frac{-2p}{2}, \frac{1}{2}, \frac{-\alpha \, s(k)}{2\sigma^2}\right) +$$

$$\sum_{k_1 < k_2} \left(\sigma^p 2^{\frac{p}{2}} \Gamma\left(\frac{p+1}{2}\right) \frac{1}{\sqrt{\pi}}\right)^2 {}_1F_1\left(\frac{-p}{2}, \frac{1}{2}, \frac{-\alpha s(k_1)}{2\sigma^2}\right) {}_1F_1\left(\frac{-p}{2}, \frac{1}{2}, \frac{-\alpha s(k_2)}{2\sigma^2}\right) \quad (9)$$

Therefore, we can write the $E\left[y_c | H_1\right]$ as:

$$E\left[y_c | H_1\right] = \frac{\dfrac{1}{\sqrt{\pi}} \sum_{k=0}^{N-1} {}_1F_1\left(\dfrac{-p}{2}, \dfrac{1}{2}, \dfrac{-\alpha \, s^2(k)}{2\sigma^2}\right)}{\sqrt{\dfrac{1}{\sqrt{\pi}} \dfrac{\Gamma\left(\frac{2p+1}{2}\right)}{\Gamma^2\left(\frac{p+1}{2}\right)} \sum_{k=0}^{N-1} {}_1F_1\left(\dfrac{-2p}{2}, \dfrac{1}{2}, \dfrac{-\alpha \, s(k)}{2\sigma^2}\right) + \dfrac{2}{\pi} \sum_{k_1 < k_2} \Phi_{p,k_1} {}_1\Phi_{p,k_2}}} = \mathbf{g}^H \mathbf{w} \quad (10)$$

where

$$\Phi_{p,k} = {}_1F_1\left(\frac{-p}{2}, \frac{1}{2}, \frac{-1}{2}, \frac{-\alpha \, s^2(k)}{2\sigma^2}\right). \quad (11)$$

Derivation of $Var\left[y_c | H_0\right]$: For the first Hypothesis, H_0 we derive the variance value as

$$Var\left[\sum_{k=0}^{N-1} |x_i(k)|^p |H_0\right] = \sum_{k=0}^{N-1} Var\left[|v_i(k)|^p\right] = \sum_{k=0}^{N-1}\left(E\left[|v_i(k)|^{2p}\right] - \left(E\left[|v_i(k)|^p\right]\right)^2\right)$$

$$= \frac{N}{\sqrt{\pi}} \sigma^{2p} 2^p \Gamma\left(\frac{2p+1}{2}\right) - N\left(\sigma^p 2^{\frac{p}{2}} \Gamma\left(\frac{p+1}{2}\right) \frac{1}{\sqrt{\pi}}\right)^2 \quad (12)$$

Using this expression and power normalization factor(7), we can find the $Var\left[y_c | H_0\right]$ as:

$$Var\left[y_c\middle|H_0\right]=\left(\frac{\frac{N}{\sqrt{\pi}}\sigma^{2p}2^p\Gamma\left(\frac{2p+1}{2}\right)-N\left(\sigma^p2^{\frac{p}{2}}\Gamma\left(\frac{p+1}{2}\right)\frac{1}{\sqrt{\pi}}\right)^2}{\frac{N}{\sqrt{\pi}}\sigma^{2p}2^p\Gamma\left(\frac{2p+1}{2}\right)+\frac{N(N-1)}{\pi}\sigma^{2p}2^p\Gamma^2\left(\frac{p+1}{2}\right)}\right)\mathbf{w}^H\mathbf{Gw}+\partial^2\mathbf{w}^H\mathbf{w}$$

(13)

Derivation of $Var\left[y_c\middle|H_1\right]$ **:** For the second Hypothesis, H_1 we derive the variance value as

$$Var\left[\sum_{k=0}^{N-1}\left|x_i(\mathbf{k})\right|^p\middle|H_1\right]=\sum_{k=0}^{N-1}\left(E\left[\left|h_is(k)+v_i(k)\right|^{2p}\right]-\left(E\left[\left|h_is(k)+v_i(k)\right|^p\right]\right)^2\right)$$

$$=\sum_{k=0}^{N-1}\left(\frac{1}{\sqrt{\pi}}\sigma^{2p}2^p\Gamma\left(\frac{2p+1}{2}\right){}_1F_1\left(-p,\frac{1}{2},\frac{-\alpha s(k)}{2\sigma^2}\right)-\left(\frac{1}{\sqrt{\pi}}\sigma^p2^{\frac{p}{2}}\Gamma\left(\frac{p+1}{2}\right){}_1F_1\left(-\frac{p}{2},\frac{1}{2},\frac{-\alpha s(k)}{2\sigma^2}\right)\right)^2\right)$$

(14)

Using the power normalization factor in Eqns.(9), (14), and (11), we obtain

$$Var\left[y_c\middle|H_1\right]=\left(\frac{\frac{1}{\sqrt{\pi}}\sigma^{2p}2^p\left(\Gamma\left(\frac{2p+1}{2}\right)\sum_{k=0}^{N-1}\Phi_{2p,k}-\frac{1}{\sqrt{\pi}}\Gamma^2\left(\frac{p+1}{2}\right)\sum_{k=0}^{N-1}\Phi_{p,k}^2\right)}{\frac{1}{\sqrt{\pi}}\sigma^{2p}2^p\left(\sum_k\Gamma\left(\frac{2p+1}{2}\right)\Phi_{2p,k}+\sum_{k_1<k_2}\Gamma^2\left(\frac{p+1}{2}\right)\frac{1}{\sqrt{\pi}}\Phi_{p,k_1}\Phi_{p,k_2}\right)}\right)\mathbf{w}^H\mathbf{Gw}+\partial^2\mathbf{w}^H\mathbf{w}$$

(15)

6.2. Proof of Theorem 1

Under the given assumptions P_d simplifies as

$$P_d=Q\left(\frac{(A_p-B_p)\left(\frac{\beta\sqrt{PM}}{D}\right)\left(\sum_{i=1}^{D}w_l\right)+Q^{-1}(P_f)\sqrt{\left(C_p\left(\frac{\beta^2PM}{D}\right)+\delta^2\right)\left(\sum_{i=1}^{D}w_i^2\right)}}{\sqrt{\left(D_p\left(\frac{\beta^2PM}{D}\right)+\delta^2\right)\left(\sum_{i=1}^{D}w_i^2\right)}}\right)$$

This formulation of P_d shows that P_d is a function of Σw_i and Σw_i^2. However, note that P_d is independent of Σw_i. This can be shown easily by replacing $\gamma\mathbf{w}$ instead $\mathbf{w}$. Then, P_d becomes independent of γ, hence we can claim that optimal $\mathbf{w}$ is such that $\Sigma w_i=1$.Furthermore, using the above equation, we can see that the P_d is maximized when is Σw_i^2 minimized assuming $\Sigma w_i=1$. This is achieved when $\mathbf{w}=(1/D)[1\ldots1]$.

Assuming $\mathbf{w}=(1/D)[1\ldots1]$, we can take the derivative of P_d wrt.D and find the optimal D when $D\in\{1,2,\ldots,M\}$. Note that D is an integer, and one has to pay attention to the boundary of the set $\{1,2,\ldots,M\}$ while finding the D that maximized P_d. This operation will give us the optimal solution since D should be an integer.

REFERENCES

[1]   Federal Communications Commission, " Spectrum Policy Task Force," Nov. 2002.

[2]   S.Haykin, "Cognitive Radio: Brain-empowered wireless communications," IEEE transactions on Signal Processing, vol. 57, no. 9, pp. 3562-3575, Sept 2009.

[3]   J. Mitola, "Cognitive Radio: An integrated agent architecture for software defined radio," Stockholm, Sweden, 2000.

[4]   C. Stevenson, G. Chouinard, L. Zhongding , H. Wendong , S. J. Shellhammer and W. Caldwell, "IEEE 802.22: The first cognitive radio wireless regional area network standard," IEEE Communications Magazine, vol. 47, no. 1, pp. 130-138, January 2009.

[5]   S. Mishra, A. Sahai and R. Broderson, "Cooperative sensing among cognitive radios," in Proc. IEEE International Conference on Communications (ICC), Istanbul, Turkey, 2006.

[6]   J. Shen, S. Liu, L. Zeng, J. G. G. Xie and Y. Liu, "Optimisation of cooperative spectrum sensing in cognitive radio network," IET Commun, vol. 3, p. 1170–1178, Jun. 2009.

[7]   Z. Quan, S. Cui and A. Sayed, "Optimal linear cooperation for spectrum sensing in cognitive radio networks," IEEE Journal of Selected Topics in Signal Processing, vol. 2, no. 1, pp. 28-40, Feb 2008.

[8]   J. Xiao, S. Cui, Z.-Q. Luo and A. Goldsmith, "Linear coherent decentralized estimation," IEEE Transactions on Signal Processing, vol. 56, no. 2, pp. 757-770, Feb 2008.

[9]   C. Berger, M. Guerriero, S. Zhou and P. Willett, "PAC vs. MAC for decentralized detection using noncoherent modulation," IEEE Journal on Selected Areas in Communications, vol. 23, no. 2, pp. 201-220, Feb 2005.

[10]   M. Gastpar, M. Vetterli and P. Dragotti, "Sensing reality and communicating bits: A dangerous liason," IEEE Signal Processing Magazine, vol. 4, no. 23, pp. 70-83, April 2006.

[11]   M. Gatspar and M. Vetterli, "On the capacity of wireless networks: The relay case," in Proc. IEEE INFOCOM, 2002.

[12]   I. Akyildiz, B. Lo and R. Balakrishnan, "Cooperative spectrum sensing in Cognitive Radio Networks: A Survey," Physical Communications (Elsevier) Journal, vol. 4, no. 1, pp. 40-62, March 2011.

[13]   B. Sirkeci-Mergen and X. Liu, "Group-Orthogonal MAC for Cooperative spectrum sensing in Cognitive Radios," in Proc. of MILCOM, San Jose, CA, 2010.

[14]   H. Stark and J. Woods, Probability, Random Processes and Estimation Theory for Engineers, 2 ed., Upper Saddle River, NJ: Prentice-Hall Inc., 1994.

[15]   "Wolfram   Research,"   [Online].   Available: http://functions.wolfram.com/HypergeometricFunctions/Hypergeometric1F1/03/02/.

[16]   "Wikipedia," [Online]. Available: http://en.wikipedia.org/wiki/Normal_distribution.

[17]   "Wolfram   Research,"   [Online].   Available: http://functions.wolfram.com/HypergeometricFunctions/Hypergeometric1F1/03/02/05/0009/.

[18]   R.   Morelos-Zaragoza,   "Faculty   Publications,"   2007.   [Online].   Available: http://scholarworks.sjsu.edu/ee_pub/28.

[19]   W. Iqbal, "Optimal Relay Mapping for Cooperative Spectrum Sensing in Cognitive Radios," Master's Project Report, San Jose State University, San Jose, CA, May 2013.

14

ANALYTICAL MODEL FOR MOBILE USER CONNECTIVITY IN COEXISTING FEMTOCELL/MACROCELL NETWORKS

Saied M. Abd El-atty[1] and Z. M. Gharsseldien[2]

[1]Department of Computer Science and Information, Arts and Science College, Salman Bin Abdulaziz University, 54-11991,Wadi Adwassir, Kingdom of Saudi Arabia
s.soliman@sau.edu.sa
[2]Department of Mathematics, Arts and Science College, Salman Bin Abdulaziz University, 54-11991,Wadi Adwassir, Kingdom of Saudi Arabia
z.gharsseldien@sau.edu.sa

ABSTRACT

In this paper we investigate the performance of mobile user connectivity in femtocell/macrocell networks. The femto user equipment (FUE) can connect to femto access point (FAP) with low communication range rather than higher communication range to macro base station (MBS). Furthermore, in such emerging networks, the spatial reuse of resources is permissible and the transmission range can be decreased, then the probability of connectivity is high. Thereby in this study, we propose a tractable analytical model for the connectivity probability based on communication range and the mobility of mobile users in femtocell/macrocell networks. Further, we study the interplays between outage probability and spectral efficiency in such networks. Numerical results demonstrate the effectiveness of computing the connectivity probability in femtocell/macrocell networks.

KEYWORDS

Femtocell, Macrocell, Connectivity, Mobility, Communication range.

1. INTRODUCTION

Integration of femtocell technology with the existing macrocell mobile networks is a promising solution not only to improve indoor coverage but also to increase capacity of cellular mobile networks. Therefore, the deployment of femtocells technology will be useful for both users and mobile network operators, since femtocell networks introduce better quality wireless services and data transmission. Furthermore, the users make use of femtocell networks to receive strong signals, and high capacity, as well as low transmission range and power wasting [1]. As a result, the mobile network operators will have the solutions for radio resources limitations, reduction macrocell traffic load and infrastructure cost by saving lots of money by offloading most of the capital expenditure (CAPEX) and operational expenditure (OPEX) onto users [2].

In femtocell networks, the smaller size of femtocell not only provides high spectrum efficiency by using spatial reuse of resources but also decreases the transmission range and then provides high probability of connectivity.

The permanent address of the authors is
[1]The Dept. of Electronics and Electrical Communications,
Faculty of Electronic Engineering, Menoufia University, 32952, Menouf, Egypt.
[2]The Dept. of Math., Fac. Sci., Al-Azhar Uni., Nasr City,11884, Cairo, Egypt

Further, one of the most important reasons for deployment femtocells technology is to serve the remote areas with no coverage or poor signal as well as to reduce the traffic volume on the macrocell. Hence in this study, we introduce a mathematical model for the probability of connectivity as a function of communication range and the mobility of mobile user in femtocell/macrocell network. As well as, in terms of connectivity probability, we study the interplays between the outage probability and the spectral efficiency based-signal to interference ratio (SIR) in such networks.

The rest of the paper is organized as follows. The related work which discussed the technical challenges associated with femtocell deployment is presented in Section 2. Femtocell access methods comparisons are investigated in Section 3. Then, the main beneficial of femtocell technology deployment and the mathematical model for mobile user connectivity are introduced at section 4. Femtocell/macrocell network modeling and the interaction between outage probability and spectral efficiency are presented in Section 5. In Section 6, the numerical results are discussed. Finally, the paper is concluded at Section 7.

2. RELATED WORK

Most of studies and researches in Femto-Macro cellular networks are focused on proposing the schemes to overcome the technical challenges in such networks such as, interference management, handover control, spectrum allocation, access methods and etc. In [3], the authors proposed an efficient hybrid frequency assignment technique based on interference limited coverage area (ILCA). They studied two different scenarios of path loss for calculation ILCA. On the same direction the authors in [4] introduced a decentralized resource allocation scheme for shared spectrum in macro/femto OFDMA networks in order to avoid inter-cell interference. In addition, the authors in [5] exploited the femto size feature to propose a resource reuse scheme based on split reuse and graph theory. In sequel, the authors in [6] have proposed a radio resource management in self organizing femtocell and macrocell networks to satisfy the demanding of selfish users. Further, according to the proposed scheme, they have introduced incentives for femtocell users to share their FAPs with public users.

On the other hand, the authors in [7] studied the outage probability in macrocell integrated with femtocell CDMA networks. They concluded that femtocell exclusion region and a tier selection based handoff policy offers modest improvements in area spectral efficiency (ASE). The handover procedure in femtocell integrated with 3GPP LTE network is analyzed for three different scenarios: hand-in, hand-out and inter-FAP in [8]. Furthermore, the authors in [9] have proposed a novel handover decision algorithm according to the location of user in the femtocell coverage and at the same time taking into account the received signal strength (RSS) from femtocell and macrocell. In [10] the authors have investigated the existing access methods for femtocells with their benefits and drawbacks. They have also provided a description for business model and technical impact of access methods in femto/macro networks. Subsequently, a framework for femtocell access method of both licensed and unlicensed band with the coexisting with WiFi is introduced in [11]. In addition, the conflict between open and close access methods is compared in [12]. They focused on the downlink of femtocell networks and they evaluated the average throughput as a function of SINR of home and cellular users for open and closed access methods.

Unlike the above literature that mainly focus on introducing the solutions for challenges in femtocell/macrocell networks. To the best of our knowledge, our work is able to provide a convenient solution for computing the probability of connectivity in such networks. More specifically, we propose an analytical model in femtocell/macrocell networks in order to compute the probability of connectivity in terms of communication range, and mobility of mobile users. As well as we study the interplays between the probability of outage and spectral efficiency in such networks. Further, we consider a femtocellular network based-open access method (OAM) in order to enhance the macrocell users (MUEs) link reliability by selecting the closest FAPs.

3. FEMTOCELL ACCESS METHODS

3.1. Closed Access Method (CAM)

In this scenario, the FAP serves only the authorized users, therefore the macro users are not allowed to access FAP. As portrait in Fig.1, although the macro user MUE1 or MUE3 is close to the radio coverage of femtocell, they cannot access FAPs [1]. On the other hand, the femto users (FUEs) prefer CAM for private access in order to protect privacy [12]. In sequel, the femto users don't like to share the limited capacity of FAP with others if no cost revenue. However, deploying CAM causes severe cross-tier interference from macro users in the reverse link or to nearby macro users in forward link. Therefore, interference mitigation and spectrum management are technical challenges in CAM [2].

3.2. Open Access Method (OAM)

In order to reduce the cross-tier interferences the OAM is considered. In OAM any passing macro user can access a FAP if the macro user is within the radio coverage of femtocell. Alternatively the macro user MUE1 or MUE3 causes or experiences strong interference, they can access FAP as shown in Fig.2. Therefore, OAM is more efficient in improving system capacity because OAM is able to serve the macro users that causing interferences. However, OAM introduces an enormous number of handover requests and consequently high communication overhead on both the radio access networks (RAN) and IP core networks (CN) [1]. In our system model, we considered the OAM method since we would gain not only improving system capacity but also reducing interferences. On the other hand, the handover problems can be solved by handover control mechanism.

3.3. Hybrid Access Method (HAM)

As we have seen previously all access methods suffer from pros and cons. Hybrid access method (HAM) is considered an adaptive method between OAM and CAM [2]. In HAM a portion of FAP resources are reserved for exclusive use of the CAM and the remaining resources are assigned in an open manners; thereby the outage number of users in macrocell can be reduced. This procedure should be controlled to avoid high blocking probability in femtocell and to reduce the annoying of registered users [12]. Nowadays, Femto forum, Broadband forum and 3GPP specifications try to finish the most efficient access method in femtocell networks but it is still under scrutiny.

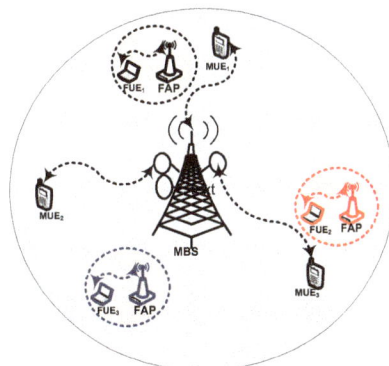

Figure 1. Closed access method (CAM) scenario.

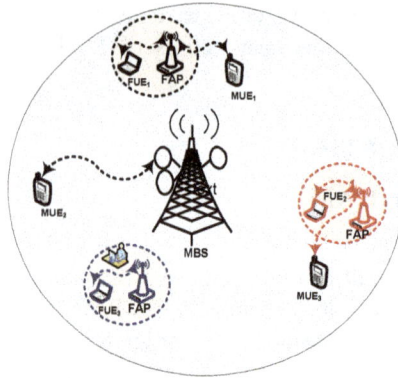

Figure 2. Closed access method (CAM) scenario.

4. BENEFITS OF DEPLOYMENT FEMTOCELLS TECHNOLOGY

Femtocells are self-organized networks (SONs) that are integrating itself into the mobile network without user intervention and then reducing deployment cost [20]. One of the key features of deployment the femtocells technology integrated with the current mobile cellular networks is that FUE or MUE requires no new equipment and hence femtocells technology does not require dual-mode handset [21]. Furthermore, the femtocell networks are capable of serving the remote areas with no coverage or poor signal as well as reducing the traffic volume on the macrocell. Therefore in terms of the communication range and the mobility factor of mobile users, in the following subsection, we introduce a tractable mathematical model for connectivity probability of mobile user.

4.1. Analytical Model for mobile user connectivity

We consider a scenario of a femtocells network; the mobile users are randomly and uniformly distributed in its coverage area. In addition, we assume a femtocells network is deployed within the communication range of a macrocell base station (MBS) and is allowed open access method (OAM) in order to enhance the mobile users (MUE) link reliability by selecting the closest FAPs.

We assume that the coverage area of femtocell is a circle with unity radius and a particular MUE with a communication range r ($r < 1$) can access FAP by employing OAM. According to the mobility of mobile users, we have three cases to study

- MUE is completely inside the femtocell range,
- MUE is completely outside the femtocell range,
- MUE communication range intersects with the femtocell range.

Let the distance between the center of FAP and the MUE is called d, which plays an important role in this analysis. There are crucial values for d that may make the MUE is completely outside of the femtocell ($d = 1 + r$ or more) or the MUE is fully inside ($d = 1 - r$ or less) as shown in Fig.3. Accordingly, we can utilize a new parameter β which expresses the area corresponding to the case of the MUE; it can be expressed as follows

$$d = 1 + \beta r \tag{1}$$

This parameter β called the mobility factor of mobile user, and plays an important role to avoid the MUE from disconnectivity.

The disconnectivity defined as the probability of at least one MUE being out of coverage region of all other femtocells in a given macrocell. The probability of MUE being in disconnectivity may be determined by calculating the area of intersection between a circle of unity radius and circle of radius r as follows:

$$A(r,\beta) = \begin{cases} 0, & \beta \geq 1 \\ h(r,\beta), & -1 < \beta < 1 \\ \pi r^2 & \beta \leq -1 \end{cases} \tag{2}$$

In this scenario, the MUE is completely outside femtocell if $\beta \geq 1$, completely inside femtocell if $\beta \leq -1$, and partially intersected with femtocell if $1 < \beta < -1$ as illustrated in Fig.3. In order to find the area of the intersection between MUE communication range's and femtocell coverage's range as a function of r and β, the equations of the circles in Cartesian (x, y) plane are given by

$$x^2 + y^2 = r^2,$$
$$(x-d)^2 + y^2 = 1 \tag{3}$$

These circles intersect at

$$x_0 = \frac{r^2 + d^2 - 1}{2d} \tag{4}$$

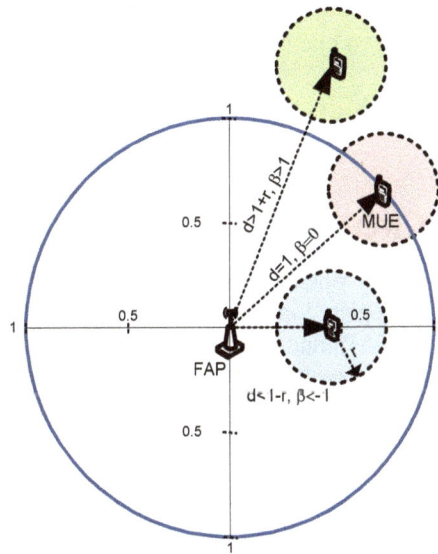

Figure 3. The mobility scenarios of mobile user in femtocell network.

By using the integration technique, we can obtain the function $h(r, \beta)$ as follows:

$$h(r,\beta) = \frac{\pi(1+r^2)}{2} - x_0\sqrt{r^2 - x_0^2}$$
$$- (d-x_0)\sqrt{1-(d-x_0)^2}$$
$$- \sin^{-1}(d-x_0) - r^2 \sin^{-1}\left(\frac{x_0}{r}\right) \tag{5}$$

All femtocells are mutually exclusive through this analysis, and each MUE only intersects with one femtocell. Hence, the intersected area with m^{th} femtocell may be written as follows:

$$A_j(r,\beta) = \begin{cases} A(r,\beta), & j=1 \\ 0 & j \neq 1 \end{cases} \tag{6}$$

Assuming that there are N_f number of femtocells overlaid in the macrocell mobile network. Hence, the probability of MUE is not able to connect to any femtocell among N_f being (P_j) is given by the probability that all N_f-1 femtocells lie in uncovered region. Then

$$P_j = \left(\frac{S - A_j(r,\beta)}{S}\right)^{N_f -1} = \left(1 - \frac{A_j(r,\beta)}{S}\right)^{N_f -1} \tag{7}$$

where $S = \pi$, j=1, 2,..., N_f is the area of the circle with unity radius for all femtocells. Thereby, the probability of at least one MUE being disconnected (P_d^1) is given by $P_d^1 = \bigcup_{x=1}^{N_f} P_x$. Upper bound on P_d^1 can be obtained by using the union bound definition as

$$P_d^1 = \sum_{x=1}^{N_f}\left(1 - \frac{A_j(r,\beta)}{S}\right)^{N_f -1} \tag{8}$$

By using (5)

$$P_d^1 \leq \left(1 - \frac{A(r,\beta)}{\pi}\right)^{N_f -1} \tag{9}$$

Thereby, we can obtain the probability of connectivity (P_c) as the complement of disconnectivity probability of at least one MUE being isolated (P_d^1), hence from (9), we obtain

$$P_C \leq \left[1 - \left(1 - \frac{A(r,\beta)}{\pi}\right)^{N_f -1}\right] \tag{10}$$

5. FEMTOCELL/MACROCELL NETWORK MODELING

We consider a network scenario of open access method (OAM) in the hierarchical macrocell with highly dense femtocells as depicted in Fig.4, where N_f femtocells are uniformly distributed in the macrocell. The density of FAPs and the density of users are represented by D_f (FAPs/m^2) and D_u (users/m^2) respectively. We assume D_u is a random stochastic Poisson process follows spatial distribution [13]. A simplified wireless channel model is considered in our network model (i.e., fading, shadowing, noise … etc. is neglected), since the channel model is a distance-dependent according to the path loss with exponent α >2. Subsequently, considering all users transmit with a fixed power P_t and all FAPs have the same power level of sensitivity, P_{min}. Thereby, we can express the communication range r in the reverse link as

$$r = \left(\frac{P_t}{P_{min}}\right)^{1/\alpha} \tag{11}$$

Furthermore, the mobile users usually try to connect to the closest FAP by sending a call request. The request may be accepted or rejected according to the available capacity in the FAP or the users are not inside the FAP coverage's range [14]. The rejected requests may be served by the overlay macrocell. Without loss of generality, we are not focusing on capacity issue in this study. We consider an OAM as access method to reduce the cross-tier interference between femtocell and macrocell usage [15]. Thereby in our analysis, the mobile user is referred to femtocell user (FUE) or macrocell user (MUE). Therefore, the probability of a given user is fully connected (P_C) to the FAP is defined as the ratio:

$$P_C = \frac{D_{f,active}}{D_u} \tag{12}$$

where $D_{f,active}$ denote to the density of active FAPs or defined as the density of active channels in a respective FAP. Therefore, P_C can be approximated as follows [14]:

$$P_C = \frac{D_f}{D_u}\left[1 - \exp\left(-\frac{D_u}{D_f}\left(1 - \exp(-D_f \cdot \pi r^2)\right)\right)\right] \tag{13}$$

Additionally, we study the effect of interference between the active FAP and mobile users in terms of outage probability [16], [17] and [18]; we consider the signal to interference (SIR) is given by:

$$SIR = \frac{P_t \times r_0^{-\alpha}}{N_0 + \sum_{i \in I} P_t \times r_i^{-\alpha}} \tag{14}$$

where r_0 represents to the distance between the reference FAP and its respective mobile user, I is the set of the interferer users, N_0 represents the noise power and r_i is the distance between the ith interferer at the reference FAP. Thereby, the outage probability (P_{outage}) is defined as the probability of FAP experiences a SIR less than or equal to a given threshold γ[19], i.e.

$$P_{outage} = \Pr\{SIR \leq \gamma\} \tag{15}$$

Hence, P_{outage} can be expressed as:

$$P_{outage} = 1 - \frac{D_f \cdot \left(1 - \exp-\left(D_{f,active}\gamma^{2/\alpha} + D_f\right) \cdot \pi r^2\right)}{\left(D_{f,active}\gamma^{2/\alpha} + D_f\right)\left[1 - \exp\left(-D_f \cdot \pi r^2\right)\right]} \tag{16}$$

The evaluation of the minimum value on the outage probability is occurred at the SIR threshold γ. Under the usual assumption in interference-limited networks, the noise power is negligible and the overall interference power is treated as Gaussian noise, hence the minimum SIR ratio γ required to guarantee a given spectral efficiency η (bits/sec. Hz) at each active link can be calculated by using the Shannon's capacity formula, which yields

$$\gamma = 2^\eta - 1 \tag{17}$$

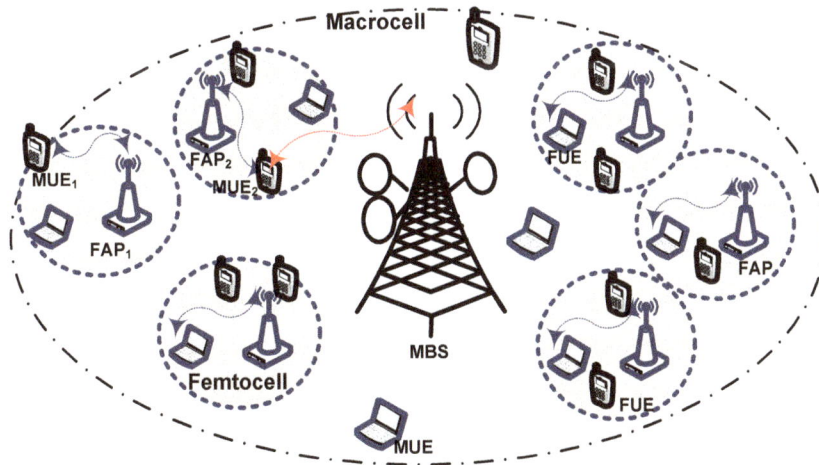

Figure 4. Femtocell/Macrocell mobile networks.

6. NUMERICAL RESULTS

In this section, we present the numerical results of mobile user connectivity in femtocell/macrocell networks. The proposed analytical model has been the basis for the implementation of network model. Different input parameters for the network model are setting according to the real cellular networks. With the aid of Mathematica packages, we obtain some numerical solution and graphical illustrations for all metrics of interest, i.e. the probability of connectivity outage probability, and spectral efficiency.

6.1. Mobile user connectivity Probability

The user connectivity probability given by (9) is function of mobility factor (β), communication range (r) and number of femtocells (N_f). We evaluate the behavior of the proposed model under two different number of femtocells $N_f = 100$ and $N_f = 10$. Fig.5 and Fig.6 illustrate the probability of mobile user connectivity. As figures indicate, the connectivity probability increases by increasing the communication range especially when the mobility factor (β) is varied from 1 towards -1, i.e., the mobile users moved from outside to inside of femtocell. In contrast, the connectivity probability decreases when the mobility factor is varied from -1 towards 1. Also, we can observe that the probability of connectivity is increased when the number of femtocells is high. This is due to when the macrocell is highly dense with femtocells, the probability of mobile user connectivity increases. In sequel, Fig.7 concludes the above discussion.

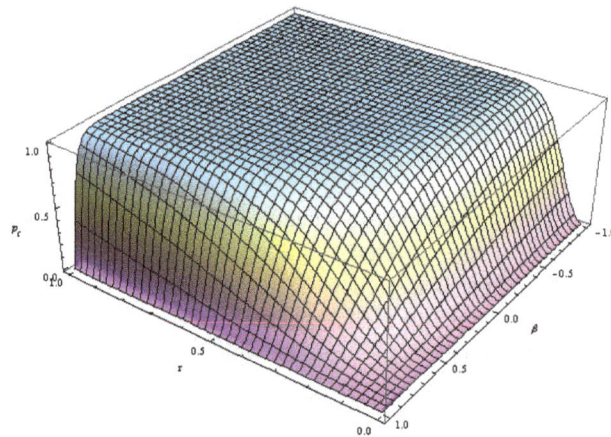

Figure 5. P_c versus β and r at $N_f = 100$.

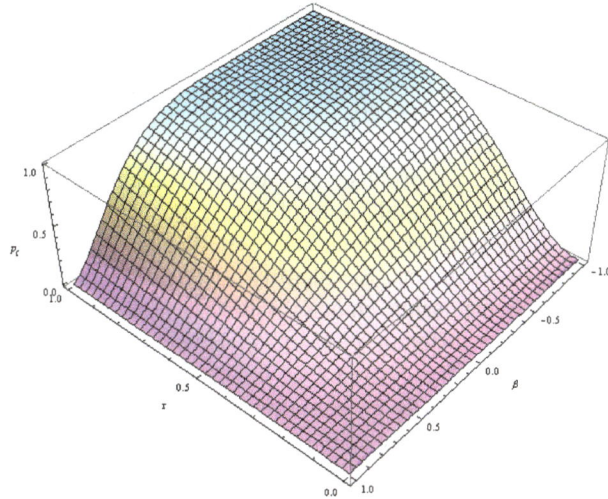

Figure 6 P_c versus β and r at N_f =10.

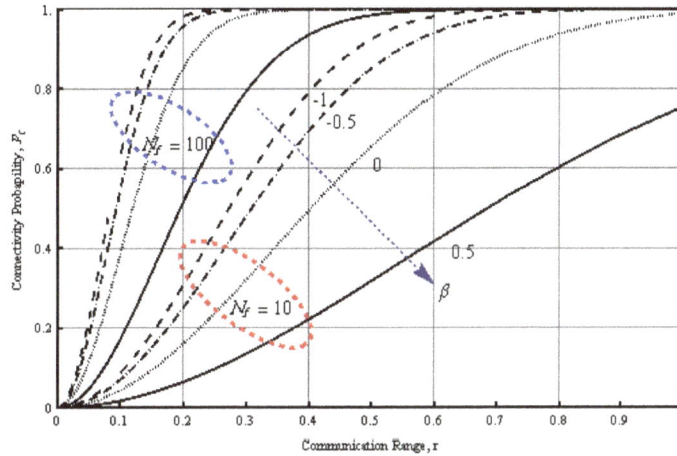

Figure 7 P_c versus r at different values of β and N_f.

6.2. Outage probability and spectral efficiency

In this section, we study the performance of the connectivity probability (P_C), outage probability (P_{outage}) and the spectral efficiency (η) at different values of femtocell density (D_f) and user density (D_u).

We present the results of P_C as a function of the density of users (D_u) at different values of femtocells densities (D_f) and at fixed communication range as shown in Fig.8. Femtocell density is produced different instances of the problem. As the figure indicates when the density of femtocells increases the user connectivity probability increases. However, when the density of users (D_u) increases, the user connectivity probability decreases. This is due to when the number of users which access the same FAP are increased, i.e., increasing the number of the rejected users, hence the user connectivity probability is decreased. In other word, the performance of femtocell/macrocell networks is significantly improved with increasing the density of femtocells.

Fig.9 shows the outage probability versus the density of femtocells at different user densities in femtocell/macrocell networks. We assumed the desired spectral efficiency in the network is

η=2bits/s.Hz and the communication range is computed as in (10) at α=4, P_r=1, P_{min}=10. Obviously, P_{out} is steadily increases at lower density of femtocells, while P_{out} decreases when the density of femtocells increased. Also, we observe P_{out} is significantly decreased after D_f=2 FAPs/m^2 for different values of users density.

On the other hand, we study the performance of spectral efficiency (η) in femtocell/macrocell networks for achieving a given threshold value of outage probability (P_{out}). Fig. 10 illustrates the spectral efficiency (η) in bits/s.Hz when the user density increasing. As expected, the spectral efficiency decreases when the user density increases, this is due to increasing the number of users that used transmission channel. Obviously, at different threshold values of outage probability, we can get different plots of η. However, the performance of spectral efficiency is increased to achieve the desired P_{out}. In other word, the plots of η are increased when the required P_{out} is high. This is probably occurred when the number of outage users increased, the channel sends with high capacity (i.e. high η) and at the same time it satisfies the desired P_{out}.

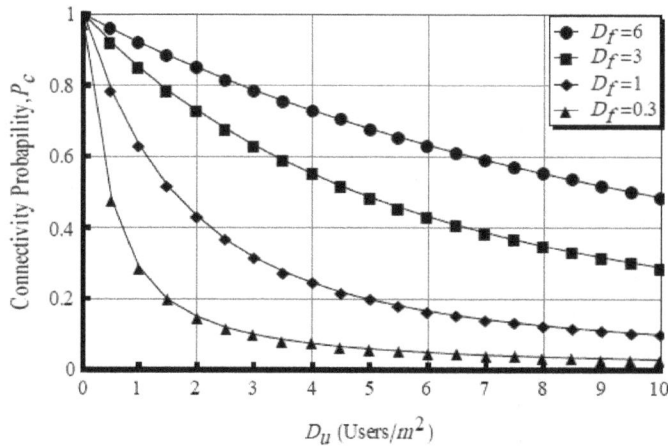

Figure 8. P_c versus D_u at different values of D_f.

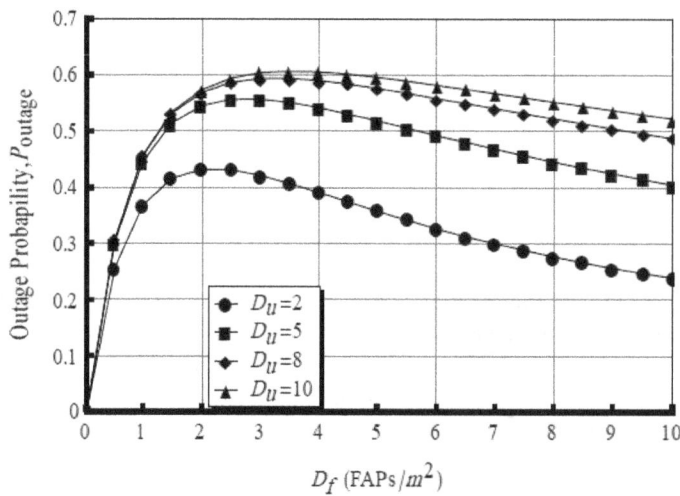

Figure 9. P_{outage} versus D_f at different values of D_u.

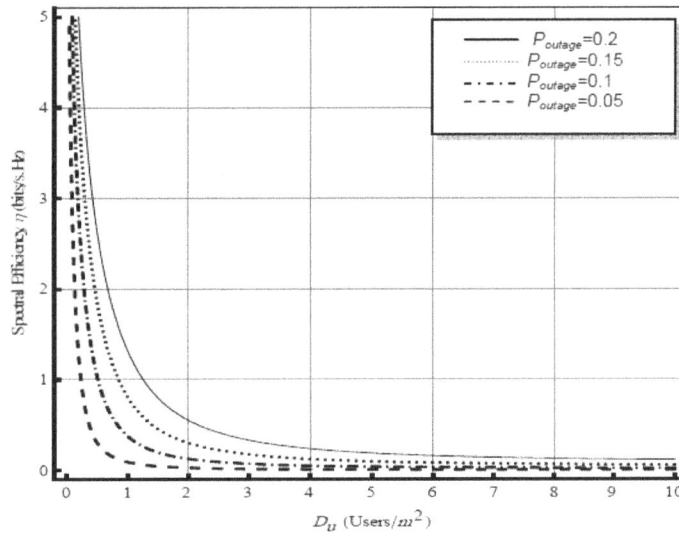

Figure 10. η versus D_u at different threshold P_{outage}.

To summarize, our analysis shows that, the probability of mobile user connectivity depends on different parameters such as communication range, femtocell density, and user density. It is shown that large number of femtocells in macrocell guaranteed feasible connectivity probability in Femto/Macro cellular networks.

7. CONCLUSIONS

This paper presented an analytical model framework for computing the probability of mobile user connectivity in femtocell/macrocell networks. The performance of the connectivity probability is measured in terms of communication range, mobility factor, user density and femtocell density. In addition, we examined the performance of outage probability and spectral efficiency in such network. Our studies demonstrated that the computing of user connectivity probability is essentially efficient during planning the Macrocellular networks integrated with Femtocellular networks.

ACKNOWLEDGMENTS

This work was supported by the Deanship of Scientific Research in Salman Bin Abdulaziz University, Kingdom of Saudi Arabia under Grants No.20/٨/1432.

REFERENCES

[1] P. Lin, J. Zhang, Y. Chen, and Q. Zhang, "Macro-Femto Heterogeneous Network Deployment and Management: From Business Models to Techinical Solutions," IEEE Wireless Comm. Magazine, Vol. 18, No.3, June. 2011, pp 64 -70.

[2] V. Chandrasekhar, J. Andrews, and A. Gatherer, "Femtocell networks: a survey," IEEE Comm. Magazine, Vol. 46, No.8, Sept. 2008, pp 59 - 67.

[3] Guvenc I., M. Jeong, Watanabe F., and Inamura H.," A hybrid frequency assignment for femtocells and coverage area analysis for co-channel operation" IEEE Comm. Letters, Vol. 12, No. 12, Dec. 2008, pp.880-882.

[4] X. Chu, Y. Wu, L. Benmesbah, and W. Kuen Ling "Resource Allocation in Hybrid Macro/Femto Networks" in Proc. IEEE WCNC Workshop, April 2010, pp. 1-6.

[5] Yongsheng S, MacKenzie A., DaSilva L.A, Ghaboosi K., and Latva-aho M.," On Resource Reuse for Cellular Networks with Femto- and Macrocell Coexistence" Proc. IEEE Globecom, Dec. 2010, pp.1-6.

[6]   C. Han Ko and H. Yu Wei, "On-Demand Resource-Sharing Mechanism Design in Two-Tier OFDMA Femtocell Networks," IEEE Trans. on Vehicular. Tech., Vol. 60, No. 3, March 2011, p.p.1059-1071.

[7]   Chandrasekhar V. and   Andrews J, "Uplink capacity and interference avoidance for two-tier femtocell networks," IEEE Trans. Wireless Comm, Vol. 8 No. 7, July 2009, pp. 3498 – 3509.

[8]   Ardian U., Robert B. and Melvi U., "Handover procedure and decision strategy in LTE-based femtocell network," Springer Telecommunication Systems, Online First, Sept. 2011.

[9]   Jung-Min M. and Dong-Ho C., "Novel Handoff Decision Algorithm in Hierarchical Macro/Femto-Cell Networks" in Proc. IEEE WCNC, April 2010, pp. 1-6.

[10]   A. Golaup, M. Mustapha, and L. Boonchin," Access control mechanisms for femtocells,"," IEEE Commun. Mag., Vol.48, No.1, January 2010, pp. 33-39.

[11]   Feilu L.,   Bala E.,   Erkip E., and   Rui Y.," A framework for femtocells to access both licensed and unlicensed bands" Proc. IEEE WiOpt, May 2011, pp.407-4011.

[12]   Han-Shin J., Ping X. and Andrews J.," Downlink Femtocell Networks: Open or Closed?" Proc. IEEE ICC, June 2011, pp.1-65.

[13]   M. Haenggi, J.G. Andrews,F. Baccelli, O. Dousse, and M. Franceschetti, "Stochastic geometry and random graphs for the analysis and design of wireless networks," IEEE Selected Area in Comm Journal, Vol. 27, No. 7, Sept. 2009, pp 1029 – 1046.

[14]   Nardelli P.H.J.,   Cardieri P. and Latva-aho, "Efficiency of Wireless Networks under Different Hopping Strategies," IEEE Trans. Wireless Comm, Vol. 11, No. 1, January 2012, pp 15 –20.

[15]   Tarasak, P.,   Quek T., and Chin F. "Uplink Timing Misalignment in Open and Closed Access OFDMA Femtocell Networks" IEEE Comm. Letters, Vol. 15, No. 9, Sept. 2011, pp. 926 – 928.

[16]   V. Chandrasekhar, M. Kountouris, and J. G. Andrews, "Coverage in multi-antenna two-tier networks," IEEE Trans. Wireless Comm.,Vol. 8, No. 10, , Oct. 2009, pp. 5314–5327

[17]   femtoforum.org/femto,"Interference Management in UMTS Femtocells," Dec. 2008.

[18]   Ngo D., Le L., Le-Ngoc T., Hossain E., and Kim D., "Distributed Interference Management in Two-Tier CDMA Femtocell Networks," IEEE Trans. Wireless Comm., In press, No.99, 2012, pp. 1-11.

[19]   K. Youngju, L. Sungeun, and H. Daesik, "Performance Analysis of Two-Tier Femtocell Networks with Outage Constraints," IEEE Trans. Wireless Comm.,Vol. 8, No. 9, Sept. 2010, pp. 5314–5327.

[20]   R. Guillaume, A. Lada, L. David, C. Chia-Chin and J. Zhang," Self-Organization for LTE Enterprise Femtocells" Proc. IEEE Globecom Workshop on Femtocell Networks, Dec. 2010, pp.674-678.

[21]   3GPP Tech. Rep. 25.820, v. 8.2.0, Sept. 2008

k-DAG Based Lifetime Aware Data Collection in Wireless Sensor Networks

Jingjing Fei[1], Hui Wu[1] and Yongxin Wang[2]

[1]School of Computer Science and Engineering, The University of New South Wales, Sydney, Australia

[2]FEIT, University of Technology, Sydney, Australia

ABSTRACT

Wireless Sensor Networks need to be organized for efficient data collection and lifetime maximization. In this paper, we propose a novel routing structure, namely k-DAG, to balance the load of the base station's neighbours while providing the worst-case latency guarantee for data collection, and a distributed algorithm for construction a k-DAG based on a SPD (Shortest Path DAG). In a k-DAG, the lengths of the longest path and the shortest path of each sensor node to the base station differ by at most k. By adding sibling edges to a SPD, our distributed algorithm allows critical nodes to have more routing choices. The simulation results show that our approach significantly outperforms the SPD-based data collection approach in both network lifetime and load balance.

KEYWORDS

Wireless sensor network; network lifetime; shortest path DAG; k-DAG; balance factor

1. INTRODUCTION

A WSN (Wireless Sensor Network) consists of a set of sensor nodes. A sensor node is composed of sensors, a processor, wireless communication components and a power module. All the senor nodes in a WSN are connected wirelessly, and work cooperatively to send the sensed data to a base station. The size of a WSN varies with applications. In a smart home, a WSN may have just dozens of sensor nodes. In a bushfire detection application, the area covered by a WSN may span several square kilometres with thousands of sensor nodes deployed. In some applications such as border surveillance, data need to be collected in real-time. Therefore, it is desirable to minimize the maximum latency of data collection, i.e., the maximum time taken by any message to arrive at the base station from the source sensor node.

In WSNs, sensor nodes are typically battery-powered, and usually deployed over a large area or in a hostile environment, which makes frequent battery replacement impractical. As a result, optimizing the energy consumption of sensor nodes is critical for extending the network lifetime. Typically, wireless communication consumes most energy of a sensor node, compared to computation and sensing [1-3]. Therefore, lowering the energy consumption of wireless communication can significantly save sensor nodes' energy, increasing the lifetime of a WSN. The communication range of a sensor node is constrained by the transmit power. To save energy, the transmit power is kept low, leading to a short transmission range. As a result, the communication between data source nodes and the base station is commonly achieved in multi-hop way. Therefore, the routing topology has a significant impact on the network lifetime.

To prolong the network lifetime, various topologies and routing algorithms have been proposed. Trees are easy to construct without much protocol overhead, and they are widely used in WSNs. In a tree, all the data converge to the base station. For each sensor node, there is only one path reaching the base station so that routing algorithms are easy to implement. However, trees are not robust enough. A link failure caused by any sensor node may isolate all its descendants from the network. Furthermore, the nodes closer to the root are more likely to die sooner as they need to relay more messages from their descendants to the root.

DAG has been proposed to improve the robustness of communication. It is more robust than a tree as each node in the network may have more than one path to the root. In addition, a DAG achieves better load balance than a tree as there are multiple paths from each source node to the base station, resulting in a longer network lifetime. Mesh network is the most robust topology. However, it induces more intricate routing algorithms than a simple tree.

In this paper, we study the problem of lifetime and latency aware data collection in a static WSN where the locations of all the sensor nodes are fixed and there is only one base station. Our objective is to maximize the network lifetime while providing the worst-case latency guarantee. The lifetime of a WSN is defined as the time when the first sensor node dies. We propose a distributed algorithm to construct a k-DAG. Our distributed algorithm constructs a k-DAG from a SPD (Shortest Path DAG) [4] by adding sibling edges. We make the following major contributions:

- We propose a novel routing structure, namely k-DAG, which can improve the lifetime and the robustness of a WSN while providing the maximum latency guarantee.

- We propose a distributed algorithm for constructing a distributed k-DAG.

- We propose a novel scheme for naming sensor nodes to support efficient point-to-point routing.

- We have simulated our approach and the approach proposed in [4]. The simulation results show that our approach outperforms theirs by up to 82% in terms of network lifetime.

- As far as we know, our approach is the first one that aims at maximizing the lifetime of a WSN while providing the maximum latency guarantee.

The rest of the paper is organized as follows. Section 2 overviews the existing approaches to lifetime aware routing. Section 3 describes our distributed algorithm for constructing a k-DAG. Section 4 presents our simulation results and analyses. Section 5 concludes the paper.

2. RELATED WORK

Lifetime aware data collection is a critical issue in WSNs. Different energy consumption models of sensor nodes have been presented and analysed, and a large number of approaches to lifetime aware data collection have been proposed.

[1] proposes a fundamental energy consumption model for sensor nodes. It considers the impacts of both the hardware and external radio environment of sensor nodes. [3] presents a realistic energy consumption model which identifies the energy consumption of each part of the sensor node and the impact of the external radio environment. The power consumption for receiving data is modelled as a constant value. For transmitting, only the power consumed by the power amplifier varies with the transmission range d while the power consumed by the other parts is a

constant. Based on the analyses and the simulation results, it shows that the single hop routing is always more energy efficient than multi-hop routing when a target is single hop reachable. This conclusion encourages the use of greedy approaches to resolve energy efficient routing issues in WSNs.

SPT (Shortest Path Tree) is a commonly used topology in WSNs as each sensor node in a SPT reaches the root with the smallest number of hops. However, a randomly constructed SPT may not increase network lifetime. [5] proposes a new weighted path cost function improving the SPT approach. In this approach, each link is assigned a weight according to its path length to the root, and a link closer to the root has a larger weight. By balancing load according to the links' weights, this approach increases network lifetime compared with those randomly constructed SPT. [6] studies the problem of finding a maximum lifetime tree from all the shortest path trees in a WSN. They first build a fat tree which contains all the shortest path trees. Then, they propose a method based on each node's number of children and its initial energy to find a minimum load shortest path tree to convert the problem into a semi-matching problem, and solve it by the min-cost max-flow approach in polynomial time. [7] proposes an approximation algorithm for maximizing network lifetime by constructing a min-max-weight spanning tree, which guarantees the bottleneck nodes having the least number of descendants. The approximation algorithm iteratively transfers some of the descendants of the nodes with the largest weight to the nodes with smaller weights.

[8] studies the load balancing problem in grid topology. It focuses on the energy consumption of the nodes which can communicate with the base station directly. As mentioned above, increasing the lifetimes of these nodes will prolong the network lifetime in most circumstances. The algorithm first builds a tree by absorbing the nodes which have the greatest load to the lightest branches to achieve the initial load balance. Then, it rebalances the tree by moving nodes from the branches with the heaviest load to the neighbouring branches with lighter load. The simulation results show that the routing trees constructed by their algorithm are more balanced than the SPT constructed by Dijkstra's algorithm.

Trees are not robust enough since each node has only one path to the base station. The topology needs to be periodically reconstructed to avoid network disconnection. SPD has been proposed to solve the robustness problem. In a SPD, each sensor node may have more than one parent. Multiple paths from each sensor to the base station increase not only robustness, but also network lifetime. [4] considers the issues of balancing the load to achieve longer network lifetime by routing on a SPD. It proposes a modified asynchronous distributed breadth-first search method that is similar to Frederickson's algorithm [9], but without the centralized synchronization between level expansions, to build a SPD. It also proposes MPE (Max-min Path Energy) and WPE (Weighted Path Energy) routing algorithms based on SPD.

[10] proposes a routing mechanism which takes advantage of siblings based on the DAG specified by Routing Protocol for Low-power and Lossy Networks (RPL) from IETF ROLL Working Group [11]. The authors present a detailed rank computation function to avoid loops in a DAG, which satisfies the policy of RPL draft. Then, they propose a routing method which allows no more than one sibling-hop per rank in the DAG to preserve the connection of the whole network while preventing loops in routing.

3. *κ*-DAG CONSTRUCTION

We aim at maximizing a WSN's lifetime by balancing the load among the base station's children as these nodes are the critical ones for network lifetime. Meanwhile, we provide the worst-case latency guarantee for message delivery. Specifically, we ensure that each message from a source sensor node v_i does not travel more than $k+D_{v_i}$ hops to reach the base station, where D_{v_i} is the minimum number of hops from v_i to the base station, and k is a fixed natural number. In this paper, the lifetime of a WSN is defined as the time when the first node depletes its energy.

3.1. Network Model

We assume that there is only one base station in the WSN. All the sensor nodes in the network are static. The wireless communication is reliable, and there is no packet loss or retransmission. All the sensor nodes in the network have the same transmission range and the same initial energy level. The base station has unlimited energy. Each sensor node generates one unit data per time unit.

We define a WSN as an undirected graph $G=(V, E)$, where V and E represent the set of sensor nodes and the set of edges denoting communication links, respectively. There are n sensor nodes in the WSN. Each sensor node is denoted by v_i ($i=1, 2, …, n$). Especially, v_0 denotes the base station. An edge $e_{ij}=(v_i, v_j)$ exists in E only if v_i and v_j can communicate with each other directly. The graph G is called connectivity graph. We assume that G is connected. Each sensor node has no knowledge of other sensor nodes in the network at the network initiation stage.
A spanning DAG of G is a DAG for data collection satisfying the following constraints:

● The base station is the only source node.

● Each sensor node sends its data only to its parents.

● For each sensor node v_i, there is a directed path from the base station to v_i.

A SPD is a spanning DAG of G such that for each sensor node v_i, each path from the base station to v_i is a shortest path.

A k-DAG is a spanning DAG of G such that for each sensor node v_i, the lengths of any two paths from the base station to v_i differ by at most k.

Given a spanning DAG and a sensor node v_i, the DAG rooted at v_i is a subgraph of the spanning DAG where the set of nodes includes v_i and all the nodes reachable from v_i, and the set of edges contains all the edges reachable from v_i.

3.2. SPD and SPT Constructions

Our approach needs to construct a SPD and a SPT at the beginning. The SPD is used to construct a k-DAG, and the SPT is used for efficient point-point communication.

A SPD can be constructed by using the algorithm proposed in [4] which employs the relaxation technique proposed in [12]. A SPT can be constructed from a SPD by selecting only one parent for each sensor node.

3.3. Naming

We propose a distributed naming algorithm to assign a unique ID to each sensor node. With these IDs, the base station is able to send a message to any node without flooding. The naming is based on a SPT of the network. The ID of each sensor node is a natural number between 1 and n, where n is the total number of sensor nodes in the network. The ID of the base station is 0.

Given a subtree T and a set of consecutive natural numbers between m and $m+size(T)-1$, where m is a natural number and $size(T)$ is the number of nodes in the tree T, the ID of each node in T is recursively defined as follows.

- The ID of the root of T is m.

- F or each child v_i ($i=1, 2, ..., k$) of the root of T, the ID of each sensor node in the subtree rooted at v_i is a natural number between m_i and $m_i+size(T_i)-1$, where $size(T_i)$ is the number of nodes in the subtree T_i rooted at v_i, and m_i is defined as follows.

  1. m_1 is equal to m.

  2. For each i ($i>1$), m_i is equal to $m + \sum_{j=1}^{i-1} size(T_j)$.

Intuitively, the ID of each sensor node is its rank in the depth-first search order of the SPT. However, distributed depth-first search is slow. The above definition underpins a faster distributed algorithm for implementing our naming scheme.

Our distributed naming algorithm consists of three phases. In the first phase, the base station initiates a message informing each sensor node v_i to compute the size of the subtree rooted at v_i. This message is sent to each sensor node in the network. In the second phase, starting from the leaf nodes, each sensor node v_i calculates the size of the subtree rooted at v_i after receiving the sizes of the subtrees rooted at its children. In the third phase, starting from the children of the base station, each sensor node assigns a unique ID to itself. Our algorithm uses the following messages.

- CALCULATE-SUBTREE-SIZE(v_i). This message is used to inform each sensor node v_i to calculate the size of the subtree rooted at v_i.

- SUBTREE-SIZE(v_i, $size_i$). After each sensor node v_i calculates the size $size_i$ of the subtree rooted at v_i, it sends this message to its parent in the SPT.

- ASSIGN-ID(v_i, $min\text{-}id$, $max\text{-}id$). This message is initiated by the base station and sent to each sensor node v_i. $min\text{-}id$ and $max\text{-}id$ are the smallest ID and the largest ID, respectively, of all the sensor nodes in the subtree rooted at v_i.

The details of our algorithm are shown in pseudo code in Algorithm1.

Algorithm 1: Naming

For the base station v_0:

for each child v_i

    send CALCULATE-SUBTREE-SIZE(v_i) to v_i

end for

$size_0 = 0$

for each child v_i

    receive SUBTREE-SIZE(v_i, $size_i$) from v_i

end for

min-id =1

for each child v_i

    max-id= min-id + size$_i$ -1

    send ASSIGN-ID(v_i, *min-id*, *max-id*) to v_i

    min-id =max-id +1

end for

For each sensor node v_i:

receive CALCULATE-SUBTREE-SIZE(v_i) from the parent

size$_i$ =1

if v_i is a leaf node **then**

    send SUBTREE-SIZE(v_i, *size$_i$*) to the parent

    receive ASSIGN-ID(v_i, *min-id*, *max-id*) from the parent

    ID$_i$ =min-id /* The ID of v_i is min-id*/

else

    for each child v_j

        send CALCULATE-SUBTREE-SIZE(v_j) to v_j

    end for

    for each child v_j

        receive SUBTREE-SIZE(v_j, *size$_j$*) from v_j

        size$_i$ =size$_i$ +size$_j$

    end for

    send SUBTREE-SIZE(v_i, *size$_i$*) to the parent

    receive ASSIGN-ID(v_i, *min-id*, *max-id*) from the parent

    ID$_i$ =min-id

    min-id = min-id+1

    for each child v_j

        max-id= min-id + size$_j$ -1

        send ASSIGN-ID(v_j, *min-id*, *max-id*) to v_j

        min-id =max-id + 1

    end for

end if

Figure 1 shows an example of our naming scheme, where the natural number beside each sensor node is its ID.

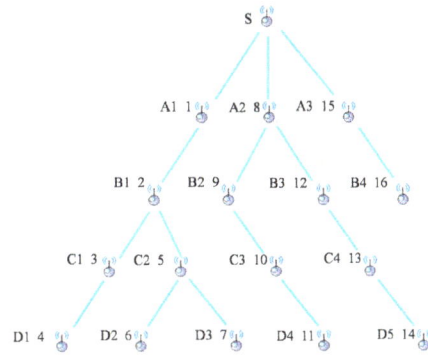

Figure 1. An example of our naming scheme

3.3. Load Calculation

After constructing a SPD and a SPT, our approach calculates the load for each sensor node. We use the definition of load in [13] to measure the data flow in the network. The load calculation is based on the DAG constructed so far. Each sensor node calculates its load as a sum of the load it produces and the load coming from its children, and distributes its load to all its parents evenly. Load calculation starts from the leaf sensor nodes in a bottom-up way and ends at the base station. After the base station collects the load from all its children, the load calculation finishes.

Algorithm 2: Load calculation

For each node v_i:

  if v_i is a leaf node **then**

    $Ld_{v_i} = 1$

    broadcast $LC(v_i, Ld_{v_i})$ to all the parents

  else if v_i is not the base station **then**

    $Ld_{v_i} = 1$

    for each child node v_j **do**

      receive $LC(v_j, load_{v_j})$ from v_j

      $Ld_{v_i} = Ld_{v_i} + load_{v_j}$

    end for

    let p_i be the number of parents of v_i

    broadcast $LC(v_i, Ld_{v_i}/p_i)$ to all the parents

  else /* v_i is the base station */

    $Ld_{v_i} = 0$

    for each child v_j **do**

      receive $LC(v_j, load_{v_j})$ from v_j

      record the load of v_j

end for

end if

We use an LC message to collect the load information:

- $LC(v_i, Ld_{v_i})$, where v_i is sender's ID, and Ld_{v_i} is the load flowing from the sender to the receiver.

Algorithm 2 describes in detail how the load of each sensor node is calculated in a distributed way. Figure 2 shows an example illustrating the process of load calculation. Leaf nodes D_1, D_2, D_3, D_4, D_5, and B_4 produce one unit data per time unit to their parents by LC message. When C_2 receive all LC messages from its children D_2 and D_3, it calculates its load which is 5/2 and sends it evenly to its parents B_1 and B_2. In this way, all the nodes send their load information to their parents, and the load converges to the base station at last. In Figure 2, the value on each edge is the load flowing through the edge.

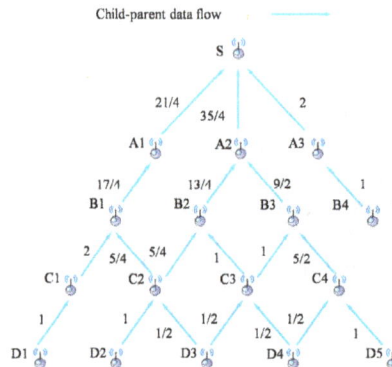

Figure 2. Load calculation based on a DAG

3.4. Adding Sibling Edges

The initial k-DAG is a SPD. After calculating the load of each child of the base station, our algorithm keeps searching for sibling edges and adding them into the k-DAG until the load balance among all the base station children is achieved. At a time, the base station finds a child v_i with the heaviest load, and a child v_j with the lightest load such that there is a sibling edge (v_t, v_s) satisfying the following constraints:

1. v_t is reachable from v_j, but not reachable from v_i.
2. v_s is reachable from v_i, but not reachable from v_j.

A sibling edge (v_t, v_s) can be added to the current k-DAG only iff the following constraints are satisfied:

3. After (v_t, v_s) is added the current k-DAG, the lengths of the longest path and the shortest path from v_s differ by at most k.
4. After (v_t, v_s) is added the current k-DAG, the new load of v_j is less than the old load of v_i.

A sibling edge (v_t, v_s) will be added if the above constraints are satisfied. Then, our algorithm tries to add adjacent sibling edges which are reachable from v_i, but not reachable from v_j before adding sibling edge (v_t, v_s). Once such an adjacent sibling edge is not available, our algorithm will go down to the next level to try to add other sibling edges. This process will be repeated until no sibling edge can be added to the current k-DAG or the current node is a leaf node.

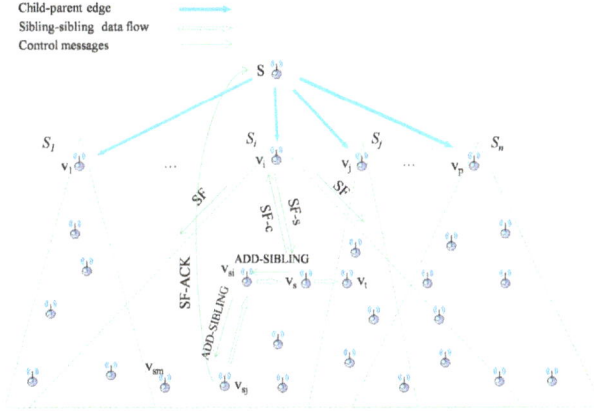

Figure 3. Adding sibling edges

We introduce the following messages:

- $SF(v_i, v_j, LdBl_{v_i})$: The base station floods this message to the DAG rooted at v_i to find a sibling edge (v_t, v_s) as discussed above. $LdBl_{v_i}$ is the ideal load that needs to be diverted from v_i to v_j. $LdBl_{v_i}$ is set to $(Ld_{v_i} - Ld_{v_j})/2$.

- $SF\text{-}c(v_s, v_i, v_j, LdC_{v_s})$: After receiving an SF message from parents, v_s will check if it is an end node of a sibling edge candidate (v_t, v_s). If it is, it will calculate the load LdC_{v_s} diverted from v_i to v_j via the sibling edge (v_t, v_s). Let p_i be the number of parents of v_s. LdC_{v_s} is equal to Ld_{v_s}/p_s.

- $SF\text{-}s(v_s, v_i, v_j, LdRe_{v_s}, SL_{v_s})$: After receiving SF-c from sibling edge candidates, the base station chooses the node v_s with the largest LdC value. The energy level of the node is used to break the tie. Then, the base station sends an SF-s message to v_s. v_s also uses SF-s to search for sibling edges. $LdRe_{v_s}$ is the remaining load that needs to be diverted from v_i to v_j, and SL_{v_s} is the number of sibling edges added to the current k-DAG.

- $ADD\text{-}SIBLING(v_p, LdRe_{v_s}, SL_{v_s})$: This message is used to inform v_p to search for a new sibling edge.

- $SF\text{-}ACK(v_{sm}, v_i, v_j, LdRe_{v_{sm}}, SL_{v_{sm}})$: v_{sm} sends SF-ACK message to the base station to indicate the end of the sibling edges search, and $LdRe_{v_{sm}}$ is the load not yet diverted from v_i to v_j.

Algorithm 3: Adding sibling edges

For the base station v_0:

exit=false

k-value=0 /* if *k-value* is equal to k, no more sibling edges can be added to the k-DAG */

while *exit=false* **do**

    find the child v_i with the heaviest load such that at least one sibling edge is added for v_i in the last round

    if such a v_i does not exists **then**

        exit=true

    exit while

    end if

    find the child v_j which is a neighbor of v_i with the lightest load

    $LdBl_{vi} = (Ld_{vi} - Ld_{vj})/2$

    broadcast $SF(v_i, v_j, LdBl_{vi})$ to v_i and all its descendants

    set timer T1

    $LdDo_{max}=0$

    repeat

        if $SF\text{-}c(v_q, v_i, v_j, LdC_{vq})$ is received from v_i's descendant v_q **then**

            if $LdC_{vq} > LdDo_{max}$ **then**

                $LdDo_{max} = LdC_{vq}$

                $v_s = v_q$

            end if

        end if

    until T1 expires

    $LdRe_{vs} = LdBl_{vi}$

    send $SF\text{-}s(v_s, v_i, v_j, LdRe_{vs}, k - k\text{-}value)$ to v_s

    loop

        if $SF\text{-}ACK(v_{sm}, v_i, v_j, LdRe_{vsm}, SL_{vsm})$ is received from v_i's descendant v_{sm} then

            $k\text{-}value = k\text{-}value - SL_{vsm}$

            broadcast a message to each sensor node in the network to recalculate its load and the reachable base station

        end if

    end loop

end while

For each sensor node v_s:

loop

if $SF(v_i, v_j, LdBl_{vi})$ is received **then**

    if this SF message is not received before **then**

        broadcast $SF(v_i, v_j, LdBl_{vi})$ to all its children

    else

        drop this SF message

    end if

    if v_s is reachable from v_j only via a sibling v_t **then**

        calculate the load LdC_{vs} diverted from v_i to v_j

        if $LdC_{vs} < LdBl_{vi}$

            send SF-c $(v_s, v_i, v_j, LdC_{vs})$ to the base station

        end if

    end if

else if $SF\text{-}s(v_s, v_i, v_j, LdRe_{vsi}, SL_{vsi})$ is received **then**

    add v_t as the parent

    $LdRe_{vs} = LdRe_{vsi} - LdC_{vs}$

    $SL_{vsi} = SL_{vsi}+1$

    if v_s has a sibling v_p and $SL_{vs} < k$ **then**

        if $LdC_{vp} \leq LdRe_{vs}$ **then**

            send ADD-SIBLING$(v_p, LdRe_{vs}, SL_{vs})$ to v_p

        end if

    else if v_s is not a leaf node and $SL_{vs} < k$ **then**

        send ADD-SIBLING$(v_p, LdRe_{vs}, SL_{vs})$ to v_p

    else

        send SF-ACK$(v_s, v_i, v_j, LdRe_{vs}, SL_{vs})$ to the base station

    end if

else if ADD-SIBLING$(v_p, LdRe_{vs}, SL_{vs})$ is received **then**

    if ADD-SIBLING$(v_p, LdRe_{vs}, SL_{vs})$ is received from a sibling **then**

        add v_p as the parent

        $LdRe_{vs} = LdRe_{vsl} - LdC_{vs}$

        $SL_{vs} = SL_{vsl}+1$

    end if

    if v_s has a sibling v_p and $SL_{vs} < k$ **then**

        if $LdC_{vp} \leq LdRe_{vs}$ **then**

            send ADD-SIBLING$(v_s, LdRe_{vs}, SL_{vs})$ to v_p

        end if

    else if v_s is not a leaf node and $SL_{vs} < k$ **then**

            send ADD-SIBLING(v_p, $LdRe_{v_s}$, SL_{v_s}) to v_p

else

            send SF-ACK(v_s, v_i, v_j, $LdRe_{v_s}$, SL_{v_s}) to the base station

       end if

    end if

end loop

Consider an example shown in Figure 3, where S_x denotes a DAG rooted at a base station's child v_x. For simplicity, we assume that k is equal to 2. First, the base station finds the child v_i with the largest load, and the child v_j with the smallest load among all its children that are also the neighbours of v_i. Next, the base station sends an SF message to all the sensor nodes in S_i. The only sibling edge is (v_s, v_t). Now, v_s sends an SF-c message to the base station. After receiving the SF-c message from the only candidate v_s, the base station sends an SF-s message to v_s. Then, v_s will add v_t as its parent, i.e., adding the sibling edge (v_s, v_t) to the k-DAG. Next, v_s sends an ADD-SIBLING(v_{si}, $LdRe_{v_s}$, SL_{v_s}) message to its sibling v_{si} to add the sibling edge (v_s,v_{si}) to the k-DAG. After that, v_{si} sends ADD-SIBLING(v_{sj}, $LdRe_{vsi}$, SL_{vsi}) to its child v_{sj}. After receiving this message, v_{sj} will not send this message to its child as no more sibling edge can be added to the k-DAG without violating the definition of the k-DAG. Therefore, v_{sj} sends SF-ACK to the base station to indicate the completion of the current round of adding sibling edges. Lastly, the base station broadcast a message to each sensor node in the network to recalculate its load and the reachable base station children.

4. SIMULATION RESULTS AND ANALYSES

We evaluate our k-DAG based approach by comparing it with the SPD based approach proposed in [4]. We use lifetime and load balance as two metrics to evaluate the performance. We implement two routing algorithms, PE and MPE proposed in [4], on these two topologies, and compare the results of these two metrics.

A sensor node's lifetime depends on its energy consumption. As mentioned in [3], the major difference for energy consumption is from transmitting and receiving. So we ignore the energy consumption for listening, computing and sensing. The initial energy of each sensor node is 0.05 J energy. Each sensor node consumes 50 nanoJ for receiving 1 bit and 250 nanoJ for sending 1 bit [14], and all the sensor nodes generate data at the rate of 40 bits/hour. The hardware platform for our simulations is Intel Core i7 processor 2.3 GHz and 8 GB RAM.

As in [8], we use Chebyshev Sum Inequality as the criteria of load balance. Let $\{v_1, v_2... v_m\}$ be the set of the base station children, and ld_{v_i} the load of a child v_i of the base station. We use the following equation to calculate the balance factor θ:

$$\theta = \frac{(\sum_{i=1}^{m} ld_{v_i})^2}{m\sum_{i=1}^{m} ld_{v_i}^2}$$

We use Cooja simulator to generate network instances, ignoring those instances with disconnected sensor nodes in the network. The transmission range for each sensor node is fixed to 50 *unit* in radius. All the sensor nodes are randomly deployed in a square area, from 100×100 to 350×350 *unit*^{2} by increasing 50×50 *unit*^{2} each time. The network size increases from 50 to 100 nodes by an increment of 10. There are a total of 6 scenarios with different network sizes, and 10 instances for each scenario, resulting in a total of 60 different network instances.

We calculate the balance factor by the data flow collected from the simulation results. The simulation results for average, maximum and minimum balance factors are shown in Figure 4, Figure 5 and Figure 6, respectively. In each figure, the horizontal axis indicates the number of sensor nodes, and the vertical axis represents the balance factor. For all the instances, *k*-DAG outperforms SPD by achieving higher load balance. The load balance improvements range from 0.1% to 83%. The largest increase of 83% occurs in a scenario with 50 sensor nodes.

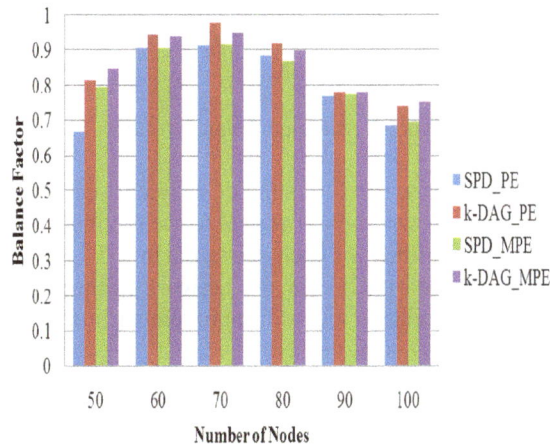

Figure 4. Average balance factors

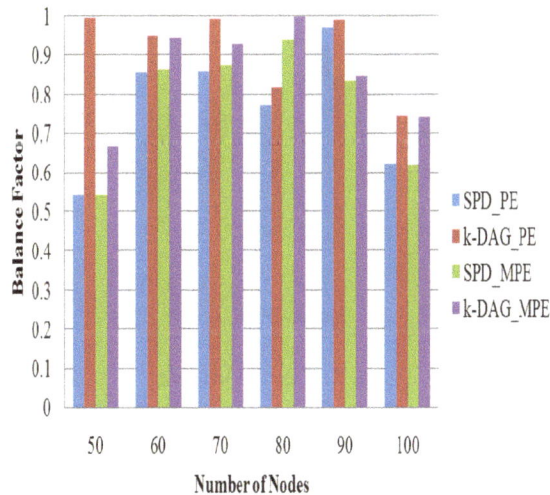

Figure 5. Maximum balance factors

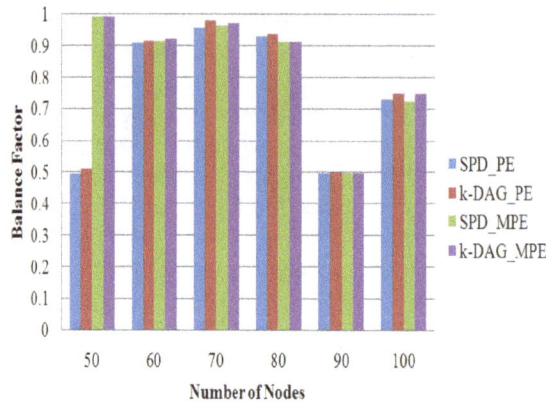

Figure 6. Minimum balance factors

Figure 7, Figure 8 and Figure 9 show the lifetimes for different scenarios. It can be seen that the lifetime of a WSN is not always inversely proportional to the number of sensor nodes of the WSN. Figure 8 shows the maximum lifetimes for each scenario. In a 50 nodes scenario, our approach achieves a maximum improvement of 82% for the network lifetime. It shows in Figure 9 that there is an instance for 50 nodes scenario with no improvement in network lifetime. However, *k-DAG* outperforms SPD for all the other instances.

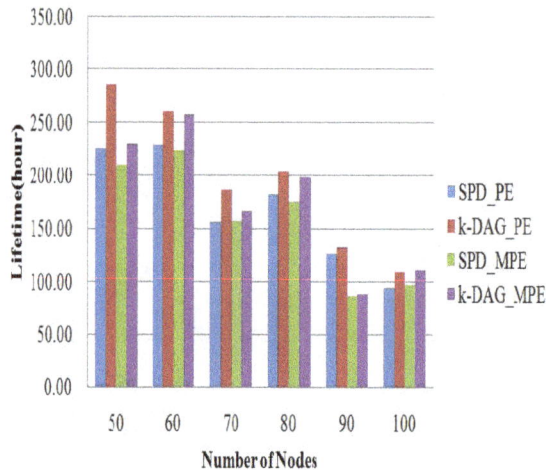

Figure 7. Average lifetimes

We choose one instance for each scenario to demonstrate the relationship between *k* and the network lifetime. In Figure 10, Max(p) is the longest path length in a *k-DAG*. The horizontal axis indicates the number of sensor nodes. The vertical axis denotes the ratio in percentage of the lifetime achieved by a particular k and the lifetime achieved by the maximum value of *k*. For 70 nodes scenario, adding the first 10% sibling edges, compared with Max(p), achieves 93.7% of the maximum lifetime. However, for a 50 nodes scenario, adding the first 10% sibling edges just achieves 15.3% of the maximum lifetime, and it improves to 70.6% when 30% sibling edges are added. It can be seen from the figure that the network lifetime is not linearly proportional to *k*. The network topology is a key factor affecting the impact of k on the network lifetime.

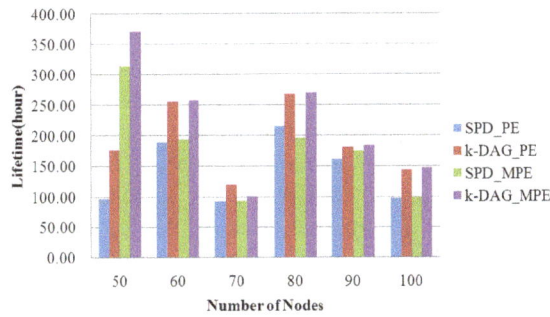

Figure 8. Maximum lifetimes

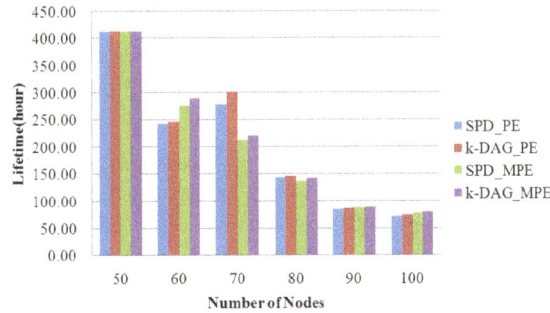

Figure 9. Minimum lifetimes

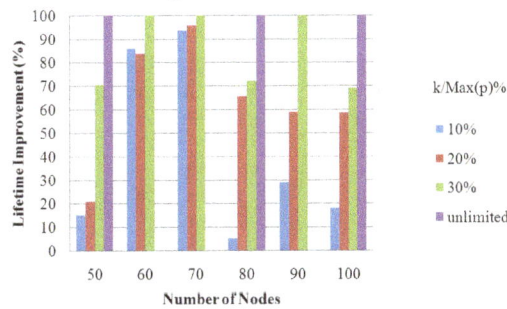

Figure 10. Lifetime versus *k*

From the simulation results, an instance with a longer network lifetime has a larger balance factor. However, a large balance factor does not guarantee a long lifetime. The number of children of the base station also has a significant impact on the network. For most instances of the same size, the more children the base station has, the longer network lifetime is achieved by our approach.

We also observe that in the instances with a small balance factor, the *k*-DAG significantly improves the network lifetime after only a few sibling edges are added into the *k*-DAG. The key reason is that the sibling edges connecting disjoint subgraphs greatly divert the load from the sensor nodes with heavy load to the sensor nodes with lighter load. Furthermore, the sibling edges at a higher level divert significantly more load than those at a lower level. In some instances, the algorithm does not optimize the network lifetime but just improves the balance factor. It occurs when the sensor node with the heaviest load among all the base station children cannot find a sibling edge to divert the load to other base station children with light load, but there are still sibling edges that can be added to the *k*-DAG for those base station children with light load. In these cases, the balance factor can be improved without lifetime increase.

5. CONCLUSION

In this paper, we study the problem of lifetime aware data collection in WSNs using DAG topology. We propose a k-DAG based approach which not only increases the lifetime of a WSN but also provides the maximum latency guarantee for data collection. We build a k-DAG in a distributed way. The k-DAG based approach achieves better load balance among the children of the base station to prolong the network lifetime. Meanwhile, it guarantees that the length of any path from each sensor node to the base station and the shortest path length differ by at most k. We have simulated our approach and compared it with the SPD based one by using a set of network instances and two routing algorithms. The simulation results show that our approach significantly outperforms the SPD based one in both network lifetime and load balance.

REFERENCES

[1]   W. R. Heinzelman, A. Chandrakasan, and H. Balakrishnan, "Energy-efficient communication protocol for wireless microsensor networks," in *System Sciences, 2000. Proceedings of the 33rd Annual Hawaii International Conference on*, 2000, p. 10 pp. vol.2.

[2]   H. Karl and A. Willig, *Protocols and architectures for wireless sensor networks*: Wiley-Interscience, 2007.

[3]   W. Qin, M. Hempstead, and W. Yang, "A Realistic Power Consumption Model for Wireless Sensor Network Devices," in *Sensor and Ad Hoc Communications and Networks, 2006. SECON '06. 2006 3rd Annual IEEE Communications Society on*, 2006, pp. 286-295.

[4]   A. Ranganathan and K. A. Berman, "Dynamic state-based routing for load balancing and efficient data gathering in wireless sensor networks," in *Collaborative Technologies and Systems (CTS), 2010 International Symposium on*, 2010, pp. 103-112.

[5]   W. Bechkit, M. Koudil, Y. Challal, A. Bouabdallah, B. Souici, and K. Benatchba, "A new weighted shortest path tree for convergecast traffic routing in WSN," in *Computers and Communications (ISCC), 2012 IEEE Symposium on*, 2012, pp. 000187-000192.

[6]   L. Dijun, Z. Xiaojun, W. Xiaobing, and C. Guihai, "Maximizing lifetime for the shortest path aggregation tree in wireless sensor networks," in *INFOCOM, 2011 Proceedings IEEE*, 2011, pp. 1566-1574.

[7]   L. Junbin, W. Jianxin, C. Jiannong, C. Jianer, and L. Mingming, "An Efficient Algorithm for Constructing Maximum lifetime Tree for Data Gathering Without Aggregation in Wireless Sensor Networks," in *INFOCOM, 2010 Proceedings IEEE*, 2010, pp. 1-5.

[8]   D. Hui and R. Han, "A node-centric load balancing algorithm for wireless sensor networks," in *Global Telecommunications Conference, 2003. GLOBECOM '03. IEEE*, 2003, pp. 548-552 Vol.1.

[9]   G. N. Frederickson, "A single source shortest path algorithm for a planar distributed network," in *STACS 85*, ed: Springer, 1985, pp. 143-150.

[10]   Q. Lampin, D. Barthel, and F. Valois, "Efficient Route Redundancy in DAG-Based Wireless Sensor Networks," in *Wireless Communications and Networking Conference (WCNC), 2010 IEEE*, 2010, pp. 1-6.

[11]   T. W. a. P. Thubert, "RPL: Routing Protocol for Low Power and Lossy Networks," 2012.

[12]   T. H. Cormen, C. E. Leiserson, R. L. Rivest, and C. Stein, *Introduction to algorithms*: MIT press, 2001.

[13]   T. Yan, Y. Bi, L. Sun, and H. Zhu, "Probability Based Dynamic Load-Balancing Tree Algorithm for Wireless Sensor Networks," in *Networking and Mobile Computing*. vol. 3619, X. Lu and W. Zhao, Eds., ed: Springer Berlin Heidelberg, 2005, pp. 682-691.

[14]   S. Mahmud and H. Wu, "Lifetime aware deployment of k base stations in WSNs," presented at the Proceedings of the 15th ACM international conference on Modeling, analysis and simulation of wireless and mobile systems, Paphos, Cyprus, 2012.

Permissions

List of Contributors

Mandeep Kaur Gondara
Ph. D Student, Computer Science Department, University of Pune, Pune

Dr. Sanjay Kadam
Research Guide, Computer Science Department, University of Pune, Pune

Natarajan Meghanathan
Jackson State University, 1400 Lynch St, Jackson, MS, USA

Vishnu Kumar Sharma
Department of CSE, JUET, Guna, Madhya Pradesh, India

Dr. Sarita Singh Bhadauria
Department of Elex, MITS Gwalior, Madhya Pradesh, India

Paramasiven Appavoo
Dept. of Computer Science & Engineering, University of Mauritius, Réduit, Mauritius

Demosthenes Vouyioukas
Department of Information and Communication Systems Engineering University of the Aegean Karlovassi 83200, Samos, Greece

Hossein Sharifi Noghabi
Department of Computer Engineering and Information Technology, Sadjad Institute of Higher Education, IRAN

Arash Ghazi askar
Department of Computer Engineering and Information Technology, Sadjad Institute of Higher Education, IRAN

Arash Boustani
Department of electrical Engineering, Wichita State University, USA

Arash Moghani
Department of Computer Engineering and Information Technology, Sadjad Institute of Higher Education, IRAN

Motahareh Bahrami Zanjani
Department of electrical Engineering, Wichita State University, USA

Rim Haddad
Laboratory Research in Telecommunication 6'Tel in High School of Communication of Tunis, Route de Raoued Km 3.5, 2083 Ariana, Tunisia

Ridha Bouallègue
Laboratory Research in Telecommunication 6'Tel in High School of Communication of Tunis, Route de Raoued Km 3.5, 2083 Ariana, Tunisia

Sultan F. Meko
IU-ATC, Department of Electrical Engineering Indian Institute of Technology Bombay, Mumbai 400 076, India

HadiTabatabaee Malazi
Department of Computer Eng., University of Isfahan, Iran

Kamran Zamanifar
Department of Computer Eng., University of Isfahan, Iran

Stefan Dulman
Embedded Software Group, Delft University of Technology, The Netherlands

Rida Khatoun
ICD - ERA - University of Technologies of Troyes (UTT), STMR, UMR CNRS 627912, rue Marie Curie 10000 - Troyes, France

Lyes Khoukhi
ICD - ERA - University of Technologies of Troyes (UTT), STMR, UMR CNRS 627912, rue Marie Curie 10000 - Troyes, France

Ahmed Nabet
ICD - ERA - University of Technologies of Troyes (UTT), STMR, UMR CNRS 627912, rue Marie Curie 10000 - Troyes, France

Dominique Gaïti
ICD - ERA - University of Technologies of Troyes (UTT), STMR, UMR CNRS 627912, rue Marie Curie 10000 - Troyes, France

T. N. Janakiraman
Department of Mathematics, National Institute of Technology, Tiruchirapalli-620015, Tamil Nadu, India

A.Senthil Thilak
Department of Mathematics, National Institute of Technology, Tiruchirapalli-620015, Tamil Nadu, India

Huda Adibah Mohd Ramli
Department of Electrical and Computer Engineering, International Islamic University Malaysia (IIUM), Kuala Lumpur, Malaysia

Kumbesan Sandrasegaran
Faculty of Engineering and Information Technology, University of Technology, Sydney, Australia

Birsen Sirkeci-Mergen
Electrical Engineering, San Jose State University, San Jose, CA

Wafa-Iqbal
Electrical Engineering, San Jose State University, San Jose, CA

Saied M. Abd El-atty
Department of Computer Science and Information, Arts and Science College, Salman Bin Abdulaziz University, 54-11991,Wadi Adwassir, Kingdom of Saudi Arabia

Z. M. Gharsseldien
Department of Mathematics, Arts and Science College, Salman Bin Abdulaziz University, 54-11991,Wadi Adwassir, Kingdom of Saudi Arabia

Jingjing Fei
School of Computer Science and Engineering, The University of New South Wales, Sydney, Australia

Hui Wu
School of Computer Science and Engineering, The University of New South Wales, Sydney, Australia

Yongxin Wang
FEIT, University of Technology, Sydney, Australia

www.ingramcontent.com/pod-product-compliance
Lightning Source LLC
Chambersburg PA
CBHW080254230326

41458CB00097B/4453